The Risk of Water Conflicts in Aotearoa-New Zealand

Adan E. Suazo

The Risk of Water Conflicts in Aotearoa-New Zealand

Emergence and Intensification

Springer

Adan E. Suazo
National Centre for Peace and Conflict Studies
University of Otago
Dunedin, New Zealand

ISBN 978-3-031-06659-7 ISBN 978-3-031-06660-3 (eBook)
https://doi.org/10.1007/978-3-031-06660-3

This Springer imprint is published by the registered company Springer Nature Switzerland AG
The registered company address is: Gewerbestrasse 11, 6330 Cham, Switzerland

Dedicated to Adán Suazo Castelar, Ismael Martinez Argueta, and the values and traditions their generation represented and defended.

Preface

Day after day, day after day,
We stuck, nor breath nor motion;
As idle as a painted ship
Upon a painted ocean.

Water, water, everywhere,
And all the boards did shrink;
Water, water, everywhere,
Nor any drop to drink.

This excerpt is part of the seminal 1834 poem 'The Rime of the Ancient Mariner' by Samuel Taylor Coleridge. It describes the anxiety attached to remaining stranded in the middle of an ocean whose waters serve no other purpose but to keep the poem's protagonists imprisoned in their ship.

A few years ago, the poem resurfaced in a rather unexpected setting. While documenting a protest in the New Zealand town of Dunedin in 2017, I noticed a poster designed by a local artist and activist that used some of the poem's lines. The poem was used to inspire tangential comparisons between its stranded ship and New Zealand's geographical isolation in the Pacific. Also, I could not help but notice that other more tacit parallels also existed, such as imagery from Māoridom like the *waka* (canoe). Above all, the poster used Coleridge's poem to describe a different kind of anxiety: one in which there is an abundance of water resources that cannot be widely enjoyed for their most fundamental use. The rally, led by Greenpeace activists, had been motivated by the intensification of water use for dairy farming in the Mackenzie River, located in the Canterbury region of New Zealand's South Island. I noted with great interest how protesters in Dunedin, a town located some 400 km south of the Mackenzie River, were willing to mobilize over environmental issues taking place elsewhere. But what seemed most noteworthy was the anxieties that drove them to protest that day.

So much of the work I had done at that point, and continue to do in the policy space today, is so eloquently represented by the above poster and by Coleridge's poem; there is water in abundance in a country like New Zealand, but anxieties over water insufficiencies and scarcities abound across the country. In New Zealand, the sense of despair felt by the mariners in Coleridge's poem can be similarly felt in the inland by several communities. The introduction of activities like water bottling has motivated significant fears for residents in New Zealand, in some cases triggering significant political mobilization. Similarly, intensive agriculture and suboptimal water infrastructure have been singled out as major sources of water contamination in several areas of the country, which have caused significant health concerns. Grievances of this nature should not be expressed in such a place of water abundance, but based on the use, manipulation and commercialization to which freshwater are subjected in New Zealand, and the myth of a water-abundant nation is constantly being challenged.

For a short period of time in the mid-1990s, and long before I would grow roots in New Zealand, my family relocated to the town of Valle de Angeles, Honduras, located some 30–40 min away from the country's capital: Tegucigalpa. Valle de Angeles, or Angels' Valley when translated in English, is a town of contrasts. While one of its main economic activities is tied to tourism, Valle de Angeles is also heavily dependent on the local trade of goods, most of which have been traditionally grown/produced in the area. Subsistence living for most of the inhabitants of this town is not a matter of choice: It is one of only a few economic enterprises in which residents can engage.

In recent years, Valle de Angeles has undergone major changes: Property value has increased drastically, highways have been built, and gas stations that were not in existence during my childhood have been installed. Some changes have arguably been conceived by the ingenuity of its people (in recent years, the town created a cooperative network of what they call 'moto-taxis', which provide an affordable and sustainable means of transportation for locals), but many others have emanated from without and with significant repercussions to local ways of life. Traditional clay pots that were once produced in local factories through centuries-old methods are increasingly being replaced by pots with the now familiar and globally recognizable label 'Made in China'.

In 2012, I returned to Valle de Angeles for the first time in many years. A local farmer shared news that a series of multinational corporations had recently been conferred mining rights in the nearby area known as San Juancito, which is also the main source of freshwater for the surrounding communities, comprised of roughly 10,000 inhabitants. I was told that because of these enterprises, violence had begun to ensue in a number of areas now owned by foreign companies, and that local groups were beginning to mobilize in protest of exploitative activities they believed would harm their livelihoods.

Shortly before my family and I moved to New Zealand in 2016, news reached us that Honduran activist Berta Caceres had been assassinated by what we now know were paramilitary groups. Caceres grew up and lived in the town of La Esperanza, Intibucá, also my mother's hometown, and home to a significant population of Indigenous Lenca. Caceres' death hit particularly hard; she was one of the most prominent voices in the environmental and Indigenous movement in Central America, and during the months leading up to her assassination, she had been highly critical of the hydroelectric dam project 'Agua Zarca'. It is believed that her participation in movements against this project contributed to her murder. Following her death, numerous acts of violence have been reported near the project.

San Juancito and 'Agua Zarca' are not stand-alone cases limited to the Honduran context; they are microcosms of incidents that are happening globally, motivated by water access and uses that are inherently incompatible with communities' aspirations and interests. Communities in Bolivia have engaged in a series of riots in the city of Cochabamba over the privatization of the city's water services. Similarly, state-mandated water diversions along the Cauvery River in India have motivated intense violence.

Episodes of water-related asymmetries are also present in Global North contexts. Canada, my second adopted country, currently faces what several observers call a 'water crisis' that consistently and disproportionately affects First Nations communities. Like New Zealand, Canada is one of the top-most water-abundant countries in the world, yet adequate water access and use are not widely beneficial across all sectors of society. Water abundance, in this regard, is not synonymous with water enjoyment for all. In view of these asymmetries, questions arise in relation to how local communities and groups may respond.

New Zealand has indeed been immune to the forms of hydropolitical violence that have plagued other nations. This however does not mean that water-motivated conflicts within its borders are absent: They simply manifest differently, and often in ways we do not fully understand. Indeed, water bottling, wholesale water marketization, intensive agricultural development, and pre-emptive water treatment have motivated communities in several parts of New Zealand to mobilize to prevent the installation of enterprises they regard as actually or potentially deleterious.

Alongside these enterprises, significant water contamination events have taken place in New Zealand, which raise questions around the current state of water infrastructure and overall management. Such is the case of the 2016 Havelock North water incident, where the presence of E. Coli in the town's water supply led to an outbreak of gastroenteritis, contributing to an estimated 5000 residents becoming ill.

On the surface, New Zealand's water abundance is such that it should satisfy the nation's multisectoral demands, and those of the ecosystems it sustains. If that is indeed the case, then why are communities in New Zealand engaging in water-related conflicts? What types of water use motivate citizens to mobilize within a water-rich environment? Why are some communities in New Zealand exposed to suboptimal water quality?

This book is an attempt to explore some of these issues.

To develop an understanding of how and why water-based conflicts emerge in a country with the environmental, economic, social, and political characteristics of New Zealand, this book poses the following question: How and why does the commercialization of freshwater affect the emergence of hydropolitical conflicts in New Zealand? To answer this question, this book proposes two central arguments: First, that water commercialization practices influence the emergence of hydropolitical conflict intentionality if they are incompatible with the interests of local communities. This argument is an acknowledgment of the varying impacts that commercial enterprises exert upon both their immediate natural environment and on the communities that draw benefits from accessing and using the resources in that environment. Therefore, it is expected that some commercial operations will be more likely to influence the emergence of conflict dynamics than others. And second, this book argues that hydropolitical conflict risk intensifies in accordance with the extent to which residents trust the local approval and appeals processes that enable water commercialization practices. This argument seeks to examine the extent to which New Zealand's water authority system can effectively pre-empt or resolve local water-based conflicts. Despite an undoubtedly robust water management infrastructure, New Zealand residents continuously engage in intense water conflicts such as protests, marches, and different types of judicial challenges, all in response to perceived inadequacies within the purview of the water authority system. The rise of intense water conflicts puts into question the country's ability to adequately pre-empt and resolve said conflicts when they intensify. This argument attempts to explain this intensification.

To elaborate the arguments presented above, this book will be divided into 10 chapters. Chapter 1 will provide an overview of the literature exploring the nexus between resource abundance and conflict. This chapter concludes that for the most part, Abundance Theory has been developed through analytical examinations of abundant non-renewable resources and has not adequately examined the conflict-causing potential of other types of abundant resources, such as freshwater. It also argues that examining freshwater abundance requires a research approach that is sensitive to the different types of values that communities attach to freshwater, which may or may not be determined by economic considerations.

Chapter 2 builds on the elaboration presented in Chap. 1 and examines New Zealand's country-specific water dynamics in the context of what is referred here as the Coleridge Dilemma. This chapter explores the quantitative state of New Zealand's freshwater supplies and presents an overview of the country's water authority framework. It later presents a brief synopsis of the types of water-based conflicts that have emerged in New Zealand, as well as the types of water-driven enterprises that have motivated these frictions.

To examine the proposed research question, Chap. 3 lays out the research design through which the proposed arguments are examined. To test the conflict-causing potential of water commercialization, this book employs three methods of analysis to examine the effects of water bottling and water chlorination in the town of Ashburton and Glenorchy, located in the Canterbury and Otago regions, respectively. First, it applies a conflict intentionality and engagement analysis, where participants' perceptions over water bottling and water chlorination are assessed against their willingness to engage in conflict over those activities, and at what level of intensity. In this regard, perceptions are argued to be shaped in relation to participants' views over the economic, environmental, and social compatibilities of water bottling and water chlorination in their towns. Second, it employs a comparative analysis to identify common causal themes across the above cases. This analysis pays close attention to the variables that influence respondents' inclination to engage in intense hydropolitical conflicts. It also endeavors to capture any potential variations in the causal impact of water bottling and water chlorination across cases. And third, this study applies a conflict intentionality classification that categorizes actors in relation to the level and type of hydropolitical conflict in which they are willing to engage, or in which they became involved.

Chapters 4 and 5 feature individual analyses as they apply to the towns of Glenorchy and Ashburton, respectively. Chapter 6 then provides a tripartite analysis of the common threads identified in both cases. The chapter concludes that the data supports the arguments presented in Chap. 3, namely that the emergence of hydropolitical conflicts in Ashburton and Glenorchy is determined by the economic, environmental and social compatibilities of water bottling and water chlorination, and that the risk of conflict intensification is predicated by the trust that residents pose upon the approval process behind each operation. However conflict was likely to escalate in the face of an untrustworthy and incompatible water operation, as was the case with water bottling in Ashburton and with water chlorination in Glenorchy, conflicts over incompatible water activities were likely to be contained within the water authority system in each location, when residents trusted the activities' approval process.

Chapter 7 builds on the findings in Chap. 6 and provides a categorization of the conflict intentionalities identified in Ashburton and Glenorchy. It categorizes parties in relation to the level of intensity of their conflict intentionality and the motivations that drove them to said potential level of engagement. In addition to this, Chap. 7 also includes an analysis of all parties that reported no visible will to engage in conflict and who chose instead to engage in Collaborative, community-based actions. While these accounts constitute a small segment of the participants, they are presented in this chapter as cases where enterprises such as water bottling and water chlorination (and in some cases, others) motivated residents to engage in positive interactions to alleviate their perceived negative impacts.

Chapter 8 explains the theoretical and empirical contributions of this study within the Abundance Theory literature, provides recommendations for future research, and describes policy intersections. This chapter stresses the need to conduct more qualitative research on the causal impacts of commercial practices such as dairy farming and to apply the study's framework in urban environments. It also calls for more research on the causal interplay between water commercialization and conflict in jurisdictions with higher concentrations of Indigenous people. Chapter 9 elaborates how a series of ongoing and proposed reforms could help address some of the issues highlighted in previous chapters. Lastly, Chap. 10 provides some final thoughts and elaborates this study's conclusions.

It must be noted that the great majority of the research, findings, and recommendations reflected in this book were written throughout a timeframe of six years. During this time, I held roles as an independent researcher and as a senior policy analyst in two government agencies. The reflections in this book are of a purely academic nature, and are in no way reflective of the New Zealand government's policy, nor do they represent the views and positions of any government department.

Lower Hutt, New Zealand — Adan E. Suazo

Acknowledgements

This book is the product of nearly six years of work, during which several people played a crucial role in supporting me as I delved deep into and engaged with the issues explored herein. My wife Ashley and daughters Amaïa and Audra-Madeleine were instrumental in me finding the time and space to explore the themes I present in this book. From defining the problems facing the water sector in New Zealand, to engaging with stakeholders and communities, my family became my haven, as they continue to be now. I am nothing but a shadow of the strength, kindness, creativity, and intelligence that my girls exhibit every day, and I am ever so grateful to learn from them every day through silent osmosis.

Through her deep intellect and literary prowess, my wife Ashley teaches me every day that the power of words can go hand in hand with beauty. The worlds she creates and the characters that inhabit them are reminders of how imagination and skillful, dedicated craft can extend the boundaries of the human mind beyond the confines of the possible. Thank you for making my stroll through life walkable. I love you.

Le debo vida, amor, protocolo y disciplina a mi madre, Marla Martínez, y a mi padre, el Embajador Adán Suazo Morazán. Este proyecto es producto de sus cuidados, consejos, y paciencia a lo largo de muchos años llenos de viajes, donde compartimos los frutos y desafíos de nuestros desplazamientos físicos, emocionales e intelectuales. Si mi padre fue responsable de la construcción del arca donde por muchos años pude reposar mi cabeza, mi madre fue el ancla que la mantuvo estable en aguas calmas y tempestuosas. No me queda tiempo en esta vida para agradecerles sus aportes en la formación de mi ser. Los amo.

This project benefited from the mentorship of several people. Dr. Katerina Standish and I first met in Ottawa in 2015 at a congress organized by the Canadian Peace Research Association, and ever since, I have been blessed with her friendship and honest advice. I have also been honored to work with her on the development of the Palgrave Handbook of Positive Peace, a journey that brought together the most outstandingly creative minds in our field. I continue to be inspired by her scholarship, her drive to advance the field, and the kindness with which she progresses her work. I know the study of peace is safe in her hands. Dr. Heather Devere and Dr. Sung Yong Lee contributed tremendously to my personal and professional development, and I always found in their counsel wisdom and empathy. I hope this book contributes to honoring the impacts they have had on me and on our field. I have also been blessed with the friendship of Maui Solomon, Susan Thorpe and Jenny Te Paa-Daniel, whose counsel and encouragement have helped me progress this work, and have allowed me the privilege of presenting my research to their communities.

My work in New Zealand was also made possible through the support of my friends and colleagues in Canada and Sweden. I want to acknowledge the support of Profs. Rosemarie Schade (Concordia University), Peter Stoett (University of Ontario Institute of Technology), Paul Shrivastava (Pennsylvania State University), Isak Svensson (Uppsala University), Ashok Swain (Uppsala University), and the late great Thomas Ohlson (Uppsala University). You all played a significant role in my personal and professional growth at different stages of my life.

Lastly, the accounts documented in this book do not belong to me; these are the histories and experiences of residents from Ashburton and Glenorchy. This book is my modest way of bringing their words back to them and to honor the time and trust they put in me throughout the data collection stages of the project.

March 2022 Adan E. Suazo

Contents

About the Author

Adan E. Suazo is a senior policy analyst at the Department of Internal Affairs (New Zealand) and a research associate at the National Centre for Peace and Conflict Studies at the University of Otago (New Zealand). His research interests include the conditions that influence the emergence and intensification of subnational water disputes and the development of stakeholder identification and engagement strategies to support environmental peacebuilding processes. His policy work has included strengthening drinking water regulations, scoping national strategies to improve stormwater management, and leading work to reconfigure local governance structures. He has previously held posts in New Zealand's Ministry for the Environment and has worked as an independent consultant on local water governance design and community engagement. As the founding coordinator of the Loyola Sustainability Research Centre at Concordia University (Canada), he worked on sustainability and peacebuilding projects with the International Secretariat for the Convention on Biological Diversity, the United Nations Environmental Education and Training Unit, and Future Earth.

Acronyms and Abbreviations

ADC	Ashburton District Council
AWA	Aotearoa Water Action
CAA	Climate Adaptation Act
CCI Scale	Conflict-Cooperation Intensity Scale
CH Model	Collier/Hoeffler Model
DWSNZ	Drinking Water Standards for New Zealand
E. Coli	*Escherichia coli*
ECan	Environment Canterbury
EIA	US Energy Information Administration
EPA	US Environmental Protection Agency
FAO	Food and Agriculture Organization
FARC-EP	Revolutionary Armed Forces of Colombia
GDP	Gross domestic product
Havelock North Report	Havelock North Drinking Water Inquiry Report
IEA	International Energy Agency
NBA	Natural and Built Environments Act
NIWA	National Institute of Water and Atmospheric Research
NNN	Nitrite-nitrate nitrogen
OECD	Organization for Economic Cooperation and Development
ORC	Otago Regional Council
QLDC	Queenstown-Lakes District Council
RMA	Resource Management Act
SPA	Strategic Planning Act
UK	United Kingdom of Great Britain and Northern Ireland
US	United States of America
USD	American Dollar
USGS	United States Geological Survey
WHO	World Health Organization
WPI	World Poverty Index

Chapter 1
Revisiting Abundance Theory

Abstract This chapter provides an overview of the literature exploring the nexus between resource abundance and conflict. It concludes that for the most part, abundance theory has been developed through analytical examinations of abundant non-renewable resources and has not adequately examined the conflict-causing potential of other types of abundant resources, such as freshwater. It also argues that examining freshwater abundance requires a research approach that is sensitive to the different types of values that communities attach to freshwater, which may or may not be determined by economic considerations.

Introduction

Abundance Theory explains how natural resource abundance can influence the emergence of conflict (Rosser, 2006). Despite a range of studies that challenge this theoretical proposition (Makdisi & Sadaka, 2006; Woodwell, 2006), there is general support for the causal effects of resource abundance on conflict. However, abundance theorists often examine this causal dyad by operationalizing natural resources as non-renewable resources. Overemphasizing the causal effects of non-renewables is problematic; this exercise is likely to lead to an empirical disregard for other types of natural resources that may bear a strong potential for conflict. Like some types of non-renewables, freshwater is regarded as a resource with an undeniably significant economic value, and several parties have been known to intensify its exploitation and manipulation globally, often causing water-related inadequacies for local communities, as well as environmental degradation and ecosystem collapse. While scholars of Neo-Malthusian persuasion have thoroughly examined how water scarcity can lead to conflict (Homer-Dixon, 1999; Hsiang et al., 2011), no comprehensive study has determined the conditions under which water conflicts may emerge in water-rich environments. To this end, this chapter will review the relevant arguments featured

Parts of this chapter were published in Suazo, A. (2021). Revisiting freshwater abundance within the peace and conflict discourse. In A. Swain et al. (eds.), *Handbook of Security and the Environment* (pp. 76–86). Cheltenham: Edward Elgar Publishing.

A. E. Suazo, *The Risk of Water Conflicts in Aotearoa-New Zealand*,
https://doi.org/10.1007/978-3-031-06660-3_1

in Abundance Theory and will present the key empirical and theoretical propositions that shape the current state of the debate.

Theoretical Arguments

In its most simple formulation, Abundance Theory suggests that 'natural resource abundance (or at least an abundance of particular types of natural resources) increases the likelihood that countries will experience negative economic, political, and social outcomes including poor economic performance, low levels of democracy, and civil war' (Rosser, 2006, p. 7). The suggestion that a so-called resource curse motivates the opening of avenues for civil unrest has attracted significant attention in the literature. This argument, as Koubi et al. note, goes against the seemingly logical proposition that natural resource wealth leads to better outcomes (2014).

From an environmental security perspective, Abundance Theory serves as a collective challenge to Neo-Malthusian Theory, whose theoretical and empirical emphasis lies on natural resource scarcity as a conflict-inducing driver (Hauge & Ellingsen, 1998; Homer-Dixon, 1999; Hsiang et al., 2011). Scholars of a Neo-Malthusian persuasion note that a reduction in natural resources is likely to produce an increasingly competitive environment, where actors will seek to maximize the gains of a depleting resource supply, potentially causing conflict over said resources. Critics of Neo-Malthusianism, known as Liberal Institutionalists, challenge its 'competition' mechanism by suggesting that water actors are more likely to seek cooperation over water (Theisen, 2012; Wolf, 2007). Applying an international relations lens to the water cooperation-conflict debate, Dinar argues that a state's ability to develop cooperative frameworks for water use depends largely on the establishment of institutions that help mitigate issues such as compliance problems and mistrust (2002). The main challenge for successful cooperation, Dinar continues, revolves around the issue of trust: 'If mistrust is alleviated and transparency enhanced, cooperation will ensue' (2002, p. 242). Dinar's argument supports Oye's observation that successful cooperation hinges upon conventions that 'create rules of thumb that [...] diminish transaction and information costs' (1986, p. 20) and on the creation of collective enforcement mechanisms (Oye, 1986). Examining the dynamics of intrastate water conflict and cooperation, Bernauer et al. explain the Liberal Institutionalist proposition by highlighting societies' adaptive capacities, which they posit include 'technological innovation, the use of the market mechanism, cooperation, and social institutions' (Bernauer et al., 2012, p. 531).

Abundance theorists have also shaped their arguments within the context of the 'New Wars' paradigm introduced by Kaldor, in which the demise of the Soviet Union, and the geopolitical dynamics that ensued, motivated the diversification of warfare economies globally, leading warring parties to rely on natural resources to sustain the war effort (Kaldor, 1999). In this context, natural resource abundance is not seen as a causing factor of conflict, but as a mechanism through which war may continue.

Empirical Elaborations

Abundance Theory literature has examined the relationship between natural resource abundance and conflict in several different ways. For instance, from an empirical perspective, the work of Collier and Hoeffler is important to note. In their quantitative examination of civil war onset, known as the Collier/Hoeffler Model (from this point on referred to as CH Model), they examine how measurements of opportunity and grievance can influence the emergence of conditions for civil war. As Collier and Hoeffler note, the outbreak of civil war requires more than just grievances: Societal discontent must be supported by some type of financing mechanism that renders war a feasible and sustainable effort (Collier & Hoeffler, 2002). Thus, by examining the economic resources available for potential rebel groups, Collier and Hoeffler argue that the onset of civil war can be predicted by analyzing jurisdictional dependence on natural resource commercialization. Of significance to this discussion is the opportunity element of the model, where indicators of natural resource dependence (measured as the ratio of primary commodity exports to gross domestic product) are used to determine the extent to which natural resource abundance impacts the onset of civil war. Indeed, their regression suggests a highly significant causal relationship.

The CH Model has been applied by Collier and Hoeffler in other studies (Collier & Hoeffler, 2004) and has set the foundations for subsequent research. For example, Rahman applies the CH Model in his analysis of the challenges facing riparian states along the Nile Basin, arguing that natural resource endowments prevent improvements in transboundary water management (2012). Similarly, Ross applies the CH Model to the separatist conflict in the Indonesian region of Aceh and concludes that the case fits with the risk factors proposed by Collier and Hoeffler, including dependence on natural resource exports (2005). In their analysis of natural resource use in the Colombian government—Revolutionary Armed Forces of Colombia (FARC-EP) conflict, Sanchez et al. conclude that the FARC-EP's involvement in the global cocaine trade contributed heavily in prolonging the duration and intensity of the war (2006).

Notwithstanding the above support for their model, Collier and Hoeffler's argument has been subjected to challenges from a range of scholars. For example, Makdisi and Sadaka note the CH Model's inability to predict the 1974 Lebanese Civil War, remarking that the likelihood of war in Lebanon, a country underwhelmingly dependent on natural resource exports, was very low during the period immediately preceding the war (2006). Similarly, Woodwell suggests that the CH Model was ill-equipped to predict the so-called Troubles conflict in Northern Ireland and notes Northern Ireland's relatively low dependence on primary exports (2006). Financing, Woodwell continues, was facilitated through involvement in illegal markets and activities, not on natural resource marketization (2006). Blum finds similar empirical discrepancies in his assessment of civil war in East Africa, where he finds evidence suggesting that the CH Model displays significant lacunas that reduce its predictive potential in an East African context (2006). Fearon challenges Collier and Hoeffler's causal mechanism (financing of rebel activities) in his analysis of oil dependence and concludes that diminished state capacities and the temptation to seize office for resource control are better predictors of civil conflict (2005). Humphreys's examination of diamond and oil abundance suggests support for Fearon's 'weak state' mechanism but concludes that poor state capacity does not fully nor singularly explain the onset of civil conflict (2005).

Since the introduction of the CH Model, several studies have expanded the literature's understanding of natural resource abundance, by examining specific types and characteristics of natural resources. For example, Ross identifies three key aspects of natural resources—lootability, obstructability, and legality—and argues that each dimension helps to facilitate different types of conflict and dictates the duration of each strife (2003). In subsequent studies, Ross finds evidence of a positive causal relationship between natural resource abundance and the likelihood of civil war, where not only its onset, but also its duration and intensity, which become aggravated by natural resource wealth (2004). Furthermore, his work assesses the impact of individual natural resources on conflict, most notably oil, diamonds, and gas (Ross, 2006).

Research has also opened the non-renewable resources 'black box' by studying how specific types of natural resources affect the likelihood of conflict. For example, Le Billon builds on the causal mechanism studied by Collier and Hoeffler, by examining how natural resources may affect not the direct onset of civil war, but states' vulnerability to civil unrest, by studying three types of resources: timber, diamonds, and oil (Le Billon, 2012). Fairhead proposes that a stronger emphasis on the political economy of conflict is required, based on the notion that most humanitarian crises occur in settings rich in natural resources (Fairhead, 2000). As in Le Billon's work, Fairhead focuses on an abundance of specific resources, which he categorizes as follows: gems, minerals, and carbons. Along similar empirical lines, subsequent studies scrutinize the individual effects of fossil fuels (de Soysa & Neumayer, 2007), diamonds (Lujala et al., 2005), and forest resources (Buhaug & Rød, 2006), collectively finding evidence that suggests a generally positive causal relationship.

The literature also examines how natural resource abundance is likely to influence conflict dynamics indirectly. For example, Sachs and Warner find evidence that countries with natural resource abundance often fail to achieve export-driven economic growth (2001). Their findings support theoretical propositions that relate economic underdevelopment and stagnation with higher probabilities of conflict (Stewart, 2002). Sachs and Warner's research aligns with the evidence in Doppelhofer et al., where measures of resource abundance (fraction of primary exports in total exports) present a negative causal relationship with economic growth (2000). Doppelhofer et al., however, find additional evidence suggesting that abundance of mineral resources increases the potential for political instability and rent-seeking behavior (2000). Subsequent analyses focusing on non-GDP-related measurements of natural resource abundance provide further support for its causal relationship with conflict (Atkinson & Hamilton, 2003; Neumayer, 2004). However, Brunnschweiler and Bulte suggest caution when examining the issue of natural resource abundance within the context of economic growth: They argue that most studies rely heavily on countries' dependence on natural resource exports and focus less on the rentability of the resource under examination (2008). Thus, they suggest that the dimensions of rentability and export dependence be scrutinized as part of a single research design (Brunnschweiler & Bulte, 2008).

The literature also frames natural resource abundance in relation to its effect on democratic performance. For example, Jensen and Wantchekon examine the path toward democratization in both resource-rich and resource-poor countries and conclude that the former group featured weaker democratic performance after the third wave of democratization than the latter (2004). They attribute this effect to vertical decision-making institutions monopolizing executive discretion over natural resource rent, which leads to an erosion in democratic development (2004). This solidifies the previous findings by Wantchekon, who concludes that natural resource abundance increases income inequality and strengthens authoritarianism (2002).

The research above elaborates how natural resource abundance has been studied by environmental security scholars, and despite a number of challenges, it lays out a generally positive causal path between resource wealth and conflict. The next section will discuss the shortcomings of the literature. It will examine the lack of focus allotted to renewable resource abundance as a causal determinant of conflict.

Discussion

There are two important elements that set the foundations for the current state of the natural resource abundance debate. Firstly, conceptions and empirical explorations of abundance rely heavily on quantitative methodologies to prove surplus, perhaps even, the non-scarcity of resources. Most of these measures come in the form of fractions of primary commodity exports (as is the case in the CH Model). And secondly, characteristics of abundance are implicitly and, I would argue, infrequently attached to their commercialization, and to their economic market value. Abundance theorists focus on the rent derived from natural resources, not just their volume. Thus, when the literature discusses the causal relationship between resource abundance and conflict, it is tacitly examining the commercialization of abundant resources with high rentability. It is crucial to establish this distinction early on, as it will illuminate the subsequent sections of this discussion, and this book.

The above considerations problematize the literature's analyses of resource abundance as a causal determinant of conflict. For the most part, the empirical evidence that supports the abundance discourse is based on examinations of non-renewable resources. Research led by Rahman (2012), Le Billon (2012), and Ross (2006), for example, is heavily dependent on the scrutiny of diamonds and fossil fuels, in the same way that Sanchez et al. (2006), Fairhead (2000), and Doppelhoffer et al. (2000) examine the effects of narcotics, carbons, and minerals, respectively. A visible exception is the work of Le Billon, who focuses on timber resources as well as diamonds and oil in his tripartite study (2012). Even though timber, an output of the forestry sector, is widely considered a renewable resource (McWhinnie, 2012), Shearman et al. challenge its renewable character, by arguing that the rapid exploitation of the world's forests has jeopardized their natural ability to regenerate their stock (2012). Notwithstanding this exception, and the somewhat unreconciled status of timber as a renewable resource, it is difficult to deny the literature's predominant focus on non-renewables.

The empirical reliance on these types of natural resources is understandable; non-renewables bear significant rentability based on their physical volume and strong market potential. For example, Goreux notes that in 1999, one kilogram of rough diamonds in Sierra Leone had a market value of $1.15 million United States Dollars (USD), which was equivalent to the yearly salary of 2000 full-time Sierra Leonean civil servants (2001). Similarly, Murphy and Acosta remark that the FARC-EP's annual revenue from its involvement in the cocaine trade in 2012 was roughly USD $1 billion (2013), with other sources citing higher return margins (Ramsey, 2012). While emphasizing the difficulties of arriving at an accurate estimate of the FARC-EP's cocaine revenue, Otis points out that all estimates represent a significant amount of financial resources as they pertain to the war effort (2014).

The empirical evidence that suggests a positive causal relationship between natural resource abundance and conflict is based on cases whose physical abundance of non-renewable resources is coupled with high economic profitability. Whether it is diamonds in Sierra Leone, narcotics in Colombia, or oil in the Sudan, the conceptualization of abundance is framed around not only quantitative measures, but also qualitative ones. The quantitative abundance of a resource is arguably less relevant if its economic value is low. A high market value for non-renewables translates into greater profitability by default for those groups who control their entry into the market. Therefore, the causal potential for conflict vis-a-vis natural resource commercialization is heavily anchored on resources whose quantitative and qualitative volumes are considered high from a market perspective.

With such profitability potential, it is unsurprising that the literature has devoted such attention to these and other resources. In providing this focus, however, the literature inadvertently universalizes the representational value of non-renewable resources and therefore fails to capture the conflict-inducing potential of other types of natural resources that may share similar abundance traits, such as freshwater. This is indeed a lacuna previously noted by Gizelis and Wooden in their examination of the direct and indirect relationships between water scarcity, governance, and conflict (2010).

Without acknowledging the implications of renewable resource abundance on the emergence of conflict, I argue that Abundance Theory displays a theoretical and empirical vacuum and requires a thorough examination of the renewable resource abundance-conflict nexus. To this end, research needs to be sensitive to case selection strategies that effectively identify renewable resource abundance within national and subnational settings. It also needs to employ methodological tools that effectively capture the different types of valuation systems attached to renewable resources. The multidimensional character of resources such as forests, for example, has been recognized by a number of organizations. In this regard, when stressing the importance of the forestry sector in West Africa, the Food and Agriculture Organization (FAO) says the following:

> The variety of cultural values and symbolic functions ascribed to the forests are as numerous and diverse as the communities and cultures of the region. Physically and mystically forests have defined the environment of communities in the region throughout time. The distinction that has been made between cultural values and the forest's functions is actually an artificial one. Tangibly and intangibly, forests feature in all aspects of culture: language, history, art, religion, medicine, politics, and even social structure itself. Forest trees may house the spirits of ancestors as well as those of the newborn. And forests are viewed in both positive and negative lights as sources of evil as well as power and munificence, as providers for, and hindrances to development (FAO, 1990, para. 1).

Similar multidimensional attributes have been assigned to resources such as freshwater. For example, eWater, an Australian organization, highlights the pluridimensionality of freshwater within Australia's Indigenous communities:

> For Indigenous people, water is an intricate part of the landscape that holds vast social, cultural and economic importance; its value is intangible. It is not easy to marry this with the quantitatively-focused western style of natural resource management which tends to separate components of the landscape into 'silos' (eWater, 2010, para. 6).

The issue of researching abundant renewable resources is therefore not a straightforward enterprise that can be approached, from a methodological and empirical perspective, and in the same way, the literature examines non-renewables. The next section elaborates this tension by focusing on freshwater and the ways in which it has been valued and how these values can conflict with one another.

The Nexus Between Water Abundance and Conflict

Freshwater provides a useful entry point for the scrutiny of renewables as causal determinants of conflict. Despite growing evidence that global water supplies are in decline (Beaumont, 1997; Gleditsch et al., 2006), freshwater remains disproportionately distributed across and within national jurisdictions, rendering water availability abundant in some locations, while remaining scarce in others (Fry, 2006). To this effect, several indexes have been developed to assess the sufficiency or scarcity of national water stocks based on their supply, demand, and distribution (Falkenmark, 1989; Sullivan, 2002). Additionally, progressive demands for food (FAO, 2009), energy (IEA, 2012), sanitation (WHO, 2017), and general consumption (EPA, 2017) continue to boost the global economic value of water.

As the rentability of freshwater resources continues to rise, so do pressures for their commercialization. The bottled water industry is a sector that has recognized the economic potential of water and has taken steps to accelerate its marketization. For example, in 2015, Transparency Market Research anticipated that the global value of the bottled water industry could rise to USD $279 billion by 2020 (2015). This prognosis was based on rising demand for bottled water. In the USA alone, the consumption of bottled water increased from 24.5 gallons per capita in 2005 to 36.5 gallons per capita in 2015, amounting to 11.7 billion gallons of water consumed in 2015, and profit margins of $14.2 billion that same year (Rodwan, 2016). The worldwide demand for bottled water products followed a strikingly similar upward pattern (Statista, 2017).

As these pressures result in progressively more intensive water exploitation and manipulation, their impacts on local ecosystems, enterprises, and inter-group dynamics have been the subject of concern. For example, Glennon framed this concern in the context of Nestlé's failed bid to bottle water from the Mecan River in rural Wisconsin. He pointed out that reducing the water supply of the Mecan, a particularly fragile ecosystem and home to a significant population of blue-ribbon trout (Glennon, 2002), by as little as 7.5 gallons, would have increased the entire waterway's temperature, impairing the natural reproductive cycles of its wildlife (Glennon, 2003). Such an event would not only affect the river's ecosystem; it would have likely had a negative impact on human livelihood. Similar concerns were raised in Ashburton (New Zealand) over the sale of a public estate with access to significant groundwater resources. The transaction was aimed at extracting an estimated 1.4 billion liters of water for bottling purposes under a 30-year permit (New Zealand Herald, 2016). After significant opposition to the sale, the project was eventually abandoned (New Zealand Herald, 2016).

An abundance of freshwater resources, as in the cases outlined above, does not guarantee that water distribution is equitable, or that freshwater supplies, from an environmental, cultural, economic, and social perspective, are administered to the benefit of communities. Barlow, for example, highlights the environmental impacts of freshwater commercialization in Australia and the USA and notes that pollution of holding lagoons in Adelaide (Australia) and boil-only advisories in Atlanta (US) and Sydney (Australia) all derive from a push to maximize the profitability of water at the expense of infrastructure investment (Barlow, 2003). De Villiers similarly remarks that, although three quarters of China's freshwater supplies are in the country's less-developed southern areas, the majority of its most water-intensive activities (industrial farming and agriculture) is found in regions with precarious volumes of water (de Villiers, 2003). This has led to significant contamination of waterways and water supplies that are unfit for human consumption and use (de Villiers, 2003). The dynamics dictating the poor state of freshwater in the above cases also frame water access and usage asymmetries in water-rich countries such as Canada, where hundreds of Indigenous communities continue to live under long-term water advisories because of fecal contamination and agricultural activities (BBC, 2018).

The above exploration prompts one important question: What do we know about water conflicts? Despite the conflict potential inherent in the commercialization of freshwater, the literature's comprehension of causal factors behind hydropolitical conflict episodes is largely based on inter-state dynamics (Wolf et al., 2003; Zeitoun & Mirumachi, 2008). In this regard, Gleick frames water conflicts in the context of international politics and argues that unavailability of freshwater is likely to generate tensions between water-poor and water-rich nations, due to the development limitations imposed by said unavailability (Gleick, 1993). Gleick later identifies four motivators that may contribute to the emergence of water conflicts: water systems as military and political goals, water as a tool of war, water as a target of war, and water resource allocation as a cause of contention (Gleick, 1998). In her study of inter-state water conflicts, Haftendorn finds evidence that suggests that conflicting water use along shared river ways can cause conflicts between the involved states (2000).

The literature has undeniably devoted significant attention to the study of water conflicts between states, but less focus has been provided to examinations of intra-state water conflicts. Notwithstanding this uneven focus, a small segment of the literature has examined the subnational dimensions of water-related conflicts. The work of Pandey, for example, illustrates the dynamics that dictate intra-state water conflicts in South Asia, which are largely determined by how freshwater resources are managed and by whom (Pandey, 2011). Pandey provides further examples of communities and groups that have engaged in conflicts over water, in situations where water availability has been poor, as was the case in areas located in India and Pakistan (Pandey, 2011). Similarly, Funder et al. investigate the emergence of hydropolitical conflict events in rural Zambia and conclude that conflictive water events are more frequent than cooperative ones because of competing water uses, including 'watering of livestock, irrigation development, and access to water for domestic uses' (2010, p. 761). The 'competing water use' mechanism identified by Funder et al. also appears in Sonderhaus and Moss' analysis of local water conflicts in Germany. Their work suggests that diverging water uses between upstream and downstream communities along the Fredersdorf Mill Stream has facilitated an escalating number of hydropolitical conflict events (2014).

Departing from questions related to surface water, research has also scrutinized the conflict-causing potential of other forms of water supplies. For example, Döring finds evidence that suggests that groundwater scarcity bears a significant potential to exacerbate communal violence (2020). Research has also examined the effect of climate-driven precipitation variation and its subsequent impacts on conflict. Even though some researchers have downplayed the causal patterns between precipitation and conflict (O'Loughlin et al., 2012), others have found evidence suggesting that human vulnerability to conflict, among other impacts, increases as precipitation patterns fluctuate (Büntgen et al., 2011; Kelley et al., 2015).

In addition to the above studies, scholarship examining intra-state water conflicts has also found evidence suggesting that cooperative outcomes are likely to emerge between water actors, despite the existence of competing uses. For example, in their quantitative analysis of the nature and scale of local water conflicts in Vietnam, Mali, Zambia, Bolivia, and Nicaragua, Ravnborg et al. suggest that local cooperation over water is more frequent but argue that conflictive water events remain consistent enough to warrant continued focus in the literature (Ravnborg et al., 2012).

In view of the existing pressures to intensify water commercialization, and the mounting economic value of water, several theoretical and empirical questions arise. While valuable in opening a theoretical ground for the study of natural resource abundance as a determinant of conflict, the literature displays a striking lack of studies that examine water-based conflicts in water-rich environments. What is known about water disputes is largely based on theories that highlight the scarcity of the resource.

In this regard and considering the understudied character of the water abundance-conflict nexus, Abundance Theory provides an important theoretical foundation upon which the above dyad can be thoroughly examined.

This book is an initial attempt to mobilize Abundance Theory to investigate water conflicts in water-wealthy environments.

References

Atkinson, G., & Hamilton, K. (2003). Savings, growth and the resource curse hypothesis. *World Development, 31*(11), 1793–1807.

Barlow, M. (2003). The world's water: A human right or a corporate good? In B. McDonald & D. Jehl (Eds.), *Whose water is it?: The Unquenchable thirst of a water-hungry world*. National Geographic Society.

BBC. (2018). Finding a solution to Canada's Indigenous water crisis. Retrieved May 16, 2019, from https://www.bbc.com/news/world-us-canada-44961490

Beaumont, P. (1997). Water and armed conflict in the middle east—fantasy or reality? In N. P. Gleditsch (Ed.), *Conflict and the environment* (pp. 355–374). Springer.

Bernauer, T., Böhmelt, T., Buhaug, H., Petter Gleditsch, N., Tribaldos, T., Berg Weibust, E., & Wischnath, G. (2012). Water-related intrastate conflict and cooperation (WARICC): A new event dataset. *International Interactions Empirical and Theoretical Research in International Relations, 38*(4), 529–545.

Blum, S. (2006). East Africa: Cycles of violence, and the paradox of peace. *ZEF discussion papers on development policy, 107*, 1–42. Retrieved April 24, 2017, from http://www.zef.de

Brunnschweiler, C. N., & Bulte, E. H. (2008). The resource curse revisited and revised: A tale of paradoxes and red herrings. *Journal of Environmental Economics and Management, 55*, 248–264.

Buhaug, H., & Rød, J. K. (2006). Local determinants of African civil wars, 1970e2001. *Political Geography, 25*, 315–335.

Büntgen, U., Tegel, W., Nicolussi, K., Mccormick, M., Frank, D., Trouet, V., Kaplan, J.O., Herzig, F., Heussner, K.U., Wanner, H., Luterbacher, J., Esper, J. (2011). 2500 years of european climate variability and human susceptibility. *Science, 331*(6017), 578–582.

Collier, P., & Hoeffler, A. (2002). On the incidence of civil war in Africa. *Journal of Conflict Resolution, 46*(1), 13–28.

Collier, P., & Hoeffler, A. (2004). Greed and grievance in civil war. *Oxford Economic Papers, 56*, 563–595.

De Soysa, I., & Neumayer, E. (2007). Resource wealth and the risk of civil war onset: Results from a new dataset of natural resource rents. *Conflict Management and Peace Science, 24*, 201–218.

De Villiers, M. (2003). Three rivers. In B. McDonald & D. Jehl (Eds.), *Whose water is it?: The unquenchable thirst of a water-hungry world*. National Geographic Society.

Dinar, S. (2002). Water, security, conflict, and cooperation. *SAIS Review, 22*(2), 229–253.

Doppelhofer, G., Miller, R. I., & Sala-I-Martin, X. (2000). *Determinants of long-term growth: A Bayesian averaging of classical estimates (Bace) Approach* (No. 266). *OECD economics department working papers*.

Döring, S. (2020). Come rain, or come wells: How access to groundwater affects communal violence. *Political Geography, 76*(January), 1–15.

EPA. (2017). *How we use water*. Retrieved May 25, 2017, from https://www.epa.gov/watersense/how-we-use-water#DailyLife

eWater. (2010). *The cultural value of water*. Retrieved October 21, 2019, from https://ewater.org.au/h2othinking/?q=2011/07/cultural-value-water

Fairhead, J. (2000). The conflict over natural and environmental resources. In E. W. Nafziger, F. Stewart, & R. Vayrynen (Eds.), *War, hunger, and displacement: Volume 1*. Oxford University Press.

Falkenmark, M. (1989). The massive water scarcity now threatening Africa: Why isn't it being addressed? *Ambio, 18*(2), 112–118.

FAO. (2009). *Global agriculture towards 2050*. Rome. Retrieved May 25, 2017, from http://www.fao.org/fileadmin/templates/wsfs/docs/Issues_papers/HLEF2050_Global_Agriculture.pdf

FAO. (1990). *The major significance of 'Minor' forest products: The local use and value of forests in the west african humid forest zone*. Retrieved October 21, 2019, from http://www.fao.org/3/t9450e/t9450e06.htm

Fearon, J. D. (2005). Primary commodity exports and civil war. *Journal of Conflict Resolution, 49*(4), 483–507.

Fry, A. (2006). *Water—Facts and Trends*. Retrieved May 10, 2017, from http://www.unwater.org/downloads/Water_facts_and_trends.pdf

Funder, M., Mweemba, C., Nyambe, I., Van Koppen, B., & Ravnborg, H. M. (2010). Understanding local water conflict and cooperation: The case of Namwala District, Zambia. *Physics and Chemistry of the Earth, 35*, 758–764.

Gizelis, T.-I., & Wooden, A. E. (2010). Water resources, institutions, & intrastate conflict. *Political Geography, 29*, 444–453.

Gleditsch, N. P., Furlong, K., Hegre, H., Lacina, B., & Owen, T. (2006). Conflicts over shared rivers: Resource scarcity or fuzzy boundaries? *Political Geography, 25*, 361–382.

Gleick, P. (1998). Water scarcity and conflict. In A. Dupont (Ed.), *The environment and security: What are the linkages?* The Australian National University.

Gleick, P. (1993). Water and conflict: Fresh water resources and international security. *International Security, 18*(1), 79–112.

Glennon, R. (2003). Bottling a birthright? In B. McDonald & D. Jehl (Eds.), *Whose water is it?: The unquenchable thirst of a water-hungry world*. National Geographic Society.

Glennon, R. (2002). The perils of groundwater pumping. *Issues in Science and Technology, 19*(1), 73–79.

Goreux, L. (2001). *Conflict diamonds* (Africa Region Working Paper Series No. 13).

Haftendorn, H. (2000). Water and international conflict. *Third World Quarterly, 21*(1), 51–68.

Hauge, W., & Ellingsen, T. (1998). Beyond environmental scarcity: causal pathways to conflict. *Journal of Peace Research, 35*(3), 299–317.

Homer-Dixon, T. (1999). *Environment, scarcity and violence*. Princeton University Press.

Hsiang, S. M., Meng, K. C., & Cane, M. A. (2011). Civil conflicts are associated with the global climate. *Nature, 476*, 438–441.

Humphreys, M. (2005). Natural resources, conflict, and conflict resolution: Uncovering the mechanisms. *Journal of Conflict Resolution, 49*(4), 508–537.

IEA. (2012). *Water for energy: Is energy becoming a thirstier resource*. Paris. Retrieved May 25, 2017, from http://www.worldenergyoutlook.org/media/weowebsite/2012/WEO_2012_Water_Excerpt.pdf

Jensen, N., & Wantchekon, L. (2004). Resource wealth and political regimes in Africa. *Comparative Political Studies, 37*(7), 816–841.

Kaldor, M. (1999). *New and old wars* (3rd ed.). Polity Press.

Kelley, C. P., Mohtadi, S., Cane, M. A., Seager, R., & Kushnir, Y. (2015). Climate change in the Fertile Crescent and implications of the recent Syrian drought. *Proceedings of the National Academy of Sciences, 112*(11), 3241–3246.

Koubi, V., Spilker, G., Zurich, E., Böhmelt, T., & Bernauer, T. (2014). Do natural resources matter for interstate and intrastate armed conflict? *Journal of Peace Research, 51*(2), 227–243.

Le Billon, P. (2012). *Wars of plunder: Conflicts, profits and the politics of resources*. Hurst & Company.

Lujala, P., Gleditsch, N. P., & Gilmore, E. (2005). A diamond curse? Civil war and a lootable resource. *Journal of Conflict Resolution, 49*(4), 538–562.

Makdisi, S., & Sadaka, R. (2006). The Lebanese civil war 1975–90. In P. Collier & N. Sambanis (Eds.), *Understanding civil war* (pp. 59–86). The World Bank.

McWhinnie, S. (2012). Renewable resource economics. *The Australian Economic Review, 45*(2), 246–254.

Murphy, H., & Acosta, L. J. (2013). *FARC controls 60 percent of drug trade—Colombia's police chief*. Retrieved May 1, 2017, from http://uk.reuters.com/article/uk-colombia-rebels-police-idUKBRE93L18Y20130422

Neumayer, E. (2004). Does the "'Resource Curse'" hold for growth in genuine income as well? *World Development, 32*(10), 1627–1640.

New Zealand Herald. (2016). *Ashburton council reneges on controversial water bottling deal.* Retrieved April 1, 2019, from https://www.nzherald.co.nz/business/news/article.cfm?c_id=3&objectid=11672314

O'Loughlin, J., Witmer, F. D. W., Linke, A. M., Laing, A., Gettelman, A., & Dudhia, J. (2012). Climate variability and conflict risk in East Africa, 1990–2009. *Proceedings of the National Academy of Sciences of the United States of America, 109*(45), 18344–18349.

Otis, J. (2014). *The FARC and Colombia's illegal drug trade.* Washington DC. Retrieved May 1, 2017 from https://www.wilsoncenter.org/sites/default/files/Otis_FARCDrugTrade2014.pdf

Oye, K. A. (1986). Explaining cooperation under Anarchy: Hypotheses and strategies. In K. A. Oye (Ed.), *Cooperation under Anarchy* (pp. 1–24). Princeton University Press.

Pandey, P. (2011). Understanding patterns of water conflicts: social and political variables. *Article South Asian Survey, 18*(1), 157–171.

Rahman, M. A. (2012). Water security: Ethiopia-Egypt transboundary challenges over the Nile River Basin. *Journal of Asian and African Studies, 48*(1), 35–46.

Ramsey, G. (2012). *FARC "Earns $2.4 to $3.5 Billion" From Drugs: Govt*. Retrieved May 1, 2017, from http://www.insightcrime.org/news-briefs/farc-earns-24-to-35-billion-drugs

Ravnborg, H. M., Bustamante, R., Cissé, A., Cold-Ravnkilde, S. M., Cossio, V., Djiré, M., Funder, M., Gómez, L.I., Le, P., Mweemba, C., Nyambe, I., Yen, N. T. B. (2012). Challenges of local water governance: The extent, nature and intensity of local water-related conflict and cooperation. *Water Policy, 14*(2), 336–357.

Rodwan, J. G. (2016). *Bottled Water 2015—Acceleration.* Retrieved March 29, 2017, from http://www.bottledwater.org/public/BWR_Jul-Aug_2016_BMC%202015%20bottled%20water%20stat%20article.pdf#overlay-context=economics/industry-statistics

Ross, M. (2006). A closer look at oil, diamonds, and civil war. *Annual Review of Political Science, 9*, 265–300.

Ross, M. (2003). Oil, drugs, and diamonds: The varying roles of natural resources in civil war. In K. Ballentine & J. Sherman (Eds.), *The political economy of armed conflict: Beyond greed & grievance.* Lynne Riener.

Ross, M. L. (2004). How do natural resources influence civil war? Evidence from thirteen cases. *International Organization, 58*, 35–67.

Ross, M. L. (2005). Resources and rebellion in Aceh, Indonesia. In P. Collier & N. Sambanis (Eds.), *Understanding civil war*. The World Bank.

Rosser, A. (2006). *The political economy of the resource curse: A literature survey.* IDS Working Paper Series, 268.

Sachs, J. D., & Warner, A. M. (2001). The curse of natural resources. *European Economic Review, 45*, 827–838.

Sanchez, F., Solimano, A., & Formisano, M. (2006). Conflict, violence, and crime in Colombia. In P. Collier & N. Sambanis (Eds.), *Understanding Civil War*. The World Bank.

Shearman, P., Bryan, J., & Laurance, W. F. (2012). Are we approaching 'peak timber' in the tropics? *Biological Conservation, 151*, 17–21.

Sondershaus, F., & Moss, T. (2014). Your resilience is my vulnerability: 'Rules in use' in a local water conflict. *Social Sciences, 3*(4), 172–192.

Statista. (2017). Bottled water consumption worldwide, 2017. Retrieved April 4, 2017, from https://www.statista.com/statistics/387255/global-bottled-water-consumption/

Stewart, F. (2002). Root causes of violent conflict in developing countries. *British Medical Journal, 324*, 342–345.

Sullivan, C. (2002). Calculating a water poverty index. *World Development, 30*(7), 1195–1210.

Theisen, O. M. (2012). Climate clashes? Weather variability, land pressure, and organized violence in Kenya. *Journal of Peace Research, 49*(1), 81–96.

Transparency Market Research. (2015). Global bottled water market is expected to reach USD 279.65 billion in 2020. By Volume, Global bottled water market is expected to reach 465.12 billion liters in 2020: Transparency Market Research. Retrieved April 5, 2017, from https://globenewswire.com/news-release/2015/01/16/697945/10115836/en/Global-Bottled-Water-Market-is-Expected-to-Reach-USD-279-65-billion-in-2020-By-Volume-Global-Bottled-Water-Market-is-Expected-to-Reach-465-12-Billion-Liters-in-2020-Transparency-Ma.html

Wantchekon, L. (2002). Why do resource abundant countries have authoritarian governments? *Journal of African Finance and Economic Development, 5*(2), 1–31.

WHO. (2017). *Financing universal water, sanitation and hygiene under the sustainable development goals.* Geneva. Retrieved May 25, 2017, from http://apps.who.int/iris/bitstream/10665/254999/1/9789241512190-eng.pdf?ua=1

Wolf, A. T. (2007). Shared waters: Conflict and cooperation. *Annual Review of Environment and Resources, 32*, 241–269.

Wolf, A. T., Yoffe, S. B., & Giordano, M. (2003). International waters: Identifying basins at risk. *Water Policy, 5*, 29–60.

Woodwell, D. (2006). The "Troubles" of Northern Ireland: Civil conflict in an economically well-developed state. In P. Collier & N. Sambanis (Eds.), *Understanding civil war*. The World Bank.

Zeitoun, M., & Mirumachi, N. (2008). Transboundary water interaction I: Reconsidering conflict and cooperation. *International Environmental Agreements: Politics, Law and Economics, 8*, 297–316.

Chapter 2
Water Abundance and Water Incompatibilities in New Zealand

Abstract This chapter examines New Zealand's country-specific water dynamics in the context of what is referred to as the Coleridge Dilemma. To do so, it explores the quantitative state of New Zealand's freshwater supplies and presents an overview of the country's water authority framework. It later presents a brief synopsis of the types of water-based conflicts that have emerged in New Zealand, as well as the kind of water-driven enterprises that have motivated these frictions.

Introduction

> It is ironic that 1000 or 2000 years ago, people living in parts of the Middle East, China, and Latin America had better systems for collecting, storing, and distributing water for human consumption and irrigation than they do today, a reflection of the high priority ancient civilizations placed on this issue. (Bourne, 1984, p. 3).

The 'high priority' to which Bourne implicitly alludes assumes a balance between social equity and environmental respect. It also relates to how human institutions and macrostructures have embraced progress by establishing increasingly intensive industries and unsustainable patterns of natural resource commodification. In the process of attaching an economic value to elements in the natural environment, freshwater has become a resource that is consistently claimed by individuals, groups, and sectors worldwide. This process sets the foundation for this book.

Bermejo conceptualizes resource commodification as 'an assertion of human control over nature and the de facto negation of the systemic character of nature' (2014, p. 22). This negation leads to a 'valuation (mercantile) and a use [of Nature] that is incompatible with the vastly complex network of life' (Bermejo, 2014, p. 23). Bermejo's argument stresses two important concerns, which should remain central from a conflict perspective: nature's highly complex and multidimensional character and the extent to which capitalism has created an environment that facilitates the

Parts of this chapter were previously published in Suazo, A. (2022). Water Abundance, Hydropolitical Disputes, and their Implications for Positive Peace in Aotearoa-New Zealand. In K. Standish et al. (eds.), *The Palgrave Handbook of Positive Peace* (pp. 797–809). Singapore: Springer Nature.

A. E. Suazo, *The Risk of Water Conflicts in Aotearoa-New Zealand*,
https://doi.org/10.1007/978-3-031-06660-3_2

superiority of one valuation (in this case of economic character) over others. Highlighting a resource's economic value and dictating access and use criteria based on said value is not only incompatible with the network of interlinkages found in nature; it also limits access to and use of water for activities and enterprises not conceived for commercial gain.

Even though freshwater abundance may theoretically satisfy society's multisectoral demands (both commercial and non-commercial), conflict may still emerge when the introduction of a commercial enterprise creates, or is perceived to create, scarcities for other parties. In this book, the emergence of said disputes will be examined in what will be referred henceforth as the Coleridge Dilemma: a context in which a political unit features water-based disputes of various intensities despite there being a volumetric abundance of water and a robust system to manage its access and use. Whereas Abundance Theory provides a wide conceptual and empirical scope for the study of resource-based conflicts, the Coleridge Dilemma alludes more specifically to the question of water-based disputes.

Considering the Coleridge Dilemma, the objective of this chapter is twofold. Firstly, it endeavors to explain the potential for hydropolitical conflicts in New Zealand by outlining the number of different claims and stakeholders involved in freshwater access and use. Because of the unique character of New Zealand's development as a binational and bicultural country, legitimate water claims are consistently asserted by a wide range of parties and sectors, making New Zealand a prime country to examine. Secondly, it explores how water conflicts may emerge as a result of water access that disproportionately favors commercial interests and practices.

The Coleridge Dilemma in New Zealand: An Assessment of Water Abundance and Management

As described above, the Coleridge Dilemma involves the development of water conflicts in water-rich settings. As such, New Zealand provides a useful backdrop for the study of said conflicts, on account of it being considered one of the most water-abundant countries in the world. Nationally, New Zealand's per capita water availability is estimated to be 107,527 m^3 per year (Gluckman, 2017), making it the fourth most water-rich country in the OECD (Statistics New Zealand, 2011). In terms of renewable water stock, New Zealand's Ministry for the Environment estimates that roughly 711 billion m^3 of water is stored in aquifers, while 320 and 440 billion m^3 is stored in lakes and rivers, respectively, (Ministry for the Environment, 2017). Abundant ground and surface water volumes are coupled with substantial amounts of precipitation, with Collins et al. estimating that New Zealand receives an average of 550 billion m^3 of rainwater every year (2015).

Water Management in New Zealand

New Zealand's wealth of water resources is coupled by a relatively robust water authority system, in the form of city/district and regional councils, which jointly govern different aspects of freshwater access, use, and overall management throughout the country. The rules and processes dictating the access and use of natural resources are clearly defined by legislation such as the 1991 Resource Management Act (RMA). The RMA is the Act of Parliament responsible for outlining the conditions for New Zealand's natural resource use and protection. The purpose of the RMA, as per the Ministry for the Environment, is:

> to ensure [that] activities like building houses, clearing bush, moving earth, taking water from a stream or burning rubbish won't harm our neighbors and our communities, or damage the air, water, soil and ecosystems that we and future generations need to survive. (Ministry for the Environment, 2009b, p. 2).

Tools enabled under the RMA include a collective of policies called national direction instruments, which stipulate the high-level objectives and minimum environmental standards that apply to specific resources. Once enacted, these policies are expected to be considered by regional councils when they make decisions over resource access and use in their regions. For example, national direction instruments that apply to freshwater bodies include the National Policy Statement for Freshwater Management 2020, the National Environmental Standards for Freshwater 2020, and the National Environmental Standard for Sources of Human Drinking Water 2007 (Ministry for the Environment, 2022a). Together, these policies are designed to provide adequate protection to freshwater bodies while enabling their safe access and use.

Along with its processes and regulations for natural resource access and use, the RMA also establishes a number of different conflict resolution mechanisms should disputes arise within its implementation. The RMA plays a tripartite role in the ventilation and settlement of hydropolitical conflicts. First, it stipulates that district and regional-level authorities are responsible for the management of the country's natural resources, including freshwater. In this regard, the RMA confers upon district councils responsibility over the effects of land use and activities on the surface of rivers and lakes, while regional councils are mandated to decide upon issues related to water quality and quantity, as well as addressing the sources of water contamination (Parliamentary Counsel Office, 1991). In delineating the jurisdictional confines of each governance level, the RMA also delineates the arbitrators responsible for the administration and delivery of conflict resolution pathways.

Second, the RMA effectively defines the process through which natural resources are accessed and used; activities can be permitted by a consenting authority, but when they are not, they typically require the granting of a resource consent by district and regional councils. A resource consent is a permission to undertake an activity that would otherwise contravene provisions under the RMA with respect to land use, subdivisions, coastal areas, beds of rivers and lakes, water, discharge of contaminants and noise (Parliamentary Counsel Office, 1991). Attached to a resource consent is an array of conditions by which a consent holder must abide to ensure that any adverse effect on the environment is appropriately managed.

A central component in the granting of a resource consent is what the RMA defines as an affected person: an individual or organization considered by a consenting authority to be adversely affected by a proposed activity (Parliamentary Counsel Office, 1991). The process to grant a resource consent involves the proper identification of and consultation with affected persons by either the council involved in the granting process or the party seeking the consent. In addition to the above approaches, councils may also start a public notification process involving advertisement of any given resource consent application in newspapers, with requests for written statements, or submissions from the general public (The Environment Court of New Zealand, 2022). In this regard, the resource consent regime considers both the environmental impacts of a proposed activity and how third parties may be affected by it.

And third, should concerns arise by affected persons during a consent application or by the general public when an application is publicly notified, the RMA dictates the procedures allowing parties to voice discontent, which include approaching their local councils with formal written submissions (The Environment Court of New Zealand, 2022). Individuals who present a submission can appear before council to elaborate their concerns, as well as holding informal prehearing meetings with the applicant to resolve any outstanding issues. Councils also have the option of referring both applicants and submitters to mediation, where conflicts can be effectively ventilated and resolved via third parties. Should decisions over a resource consent application remain unreconciled through council channels, the submitter can escalate their concerns to the Environment Court, and as a last resort, to the High Court (The Environment Court of New Zealand, 2022). As such the RMA enables the ventilation of a range of disputes, some administered locally by councils and others by centralized judicial bodies.

Alongside national direction instruments and consent granting procedures under the RMA, additional tools exist to ensure the safe access and use of freshwater. For example, until recently, the Ministry of Health administered the Drinking Water Standards for New Zealand (DWSNZ), which set out the rules that determine safety parameters for drinking water. These include quality standards, compliance criteria, reporting requirements, and remedial actions that drinking water providers, traditionally local authorities, must abide by to ensure the safe use of network water for drinking and other domestic uses (Ministry of Health, 2021). It is noted that recent legislative changes have transferred responsibilities over the DWSNZ away from the Ministry of Health and into the hands of an independent water regulator. Pending additional government decisions, the provision of drinking water services may be similarly transferred away from local authorities and become the responsibility of new multiregional water service delivery organizations. These policy changes will be further explained in Chap. 9.

Water Problems in New Zealand

Despite its enviable freshwater stock and relatively robust water management system, New Zealand has faced a series of different challenges connected to how its water resources are distributed and utilized. For example, the Water Poverty Index (WPI), which measures water access, availability, capacity to sustain access, use, and environmental factors (including quality), ranks New Zealand in the 56th position globally in great part due to substandard measurements of freshwater use for domestic, agricultural, and non-agricultural purposes (Lawrence et al., 2002).

Comparatively, advocacy for efficient water use and skepticism over water commercialization and privatization have dominated the national and development water strategies of several other water-abundant countries. For example, Gustafsson points out that several municipalities in Sweden have experimented with the transfer of water management and sewage services to the private sector, but that these ventures were readily moved back into public regimes due to comparatively higher costs for end users (Gustafsson, 2001). In terms of support for international organizations, Bond and Dugard highlight Norway's decision to divest millions of dollars from the World Bank's Public Private Investment Advisory Facility, based on the latter's support for structural adjustment programs and its inclination to commodify water in Global South countries (2008).

While efficiency and responsible state stewardship of water supplies place both Sweden and Norway in relatively high rankings in a number of indexes, doubts have been cast upon the efficient management of the resource in New Zealand. This was made evident by a recent report from New Zealand's Ministry for the Environment, which concludes that although there is robust documentation concerning approved water permits, little is known about the amounts of freshwater used in the country (Ministry for the Environment, 2017). Tanentzap et al. argue that this trend originates with New Zealand's over-reliance on regulatory mechanisms (as opposed to central government intervention) to mitigate the adverse environmental impacts of activities such as agriculture (2015), itself a direct function of market-driven resource intensification and commodification (2015).

New Zealand's experience with water management problems is not a novel phenomenon; there is robust evidence that documents a long history of bacteriological, protozoal, and chemical contamination events across the country. For example, in a report to New Zealand's Ministry of Health, Ball presents evidence of at least 25 cases of water-borne outbreaks between 1984 and 2006 (2007). In recent years, New Zealand's water-related problems have continued. Perhaps, the most salient example of these is the 2016 water-borne outbreak in the town of Havelock North, an incident in which an estimated 5000 residents became ill with gastroenteritis because of the presence of E. coli in the town's drinking water source. Havelock North motivated a thorough government investigation in the form of the Havelock North Drinking Water Inquiry, tasked with both finding the immediate sources of the outbreak, and the enabling elements in the water authority system that failed to prevent it. The inquiry recommended a suite of changes to how water is managed in New Zealand, including improvements to source water protection (Ministry for the Environment, 2022a, 2022b) and transformational reforms to the management and regulation of drinking, waste, and storm water, collectively known as Three Waters (Department of Internal Affairs, 2021). These reforms will be the subject of further analysis in Chap. 9.

The Coleridge Dilemma in New Zealand becomes evident when accounting for the growing number of water-related conflicts, motivated by how water is accessed, used, and degraded by actors and the water authority's inability to prevent its diminishment. From protests over water bottling in Canterbury (Newshub, 2016; NewstalkZB, 2017), to intense political organization over water treatment in Western Otago (Stuff, 2018) and Hawke's Bay (Newsie, 2017), disputes over water continue to generate discord.

Water access and use (commercial and otherwise) in New Zealand often display incompatibilities between the country's European/non-European (from this point on referred to as Pākehā) and Indigenous descendants (from this point on referred generically as Māori). This discrepancy emanates largely from profound differences over natural resource management, ownership, and Māori's unique constitutional standing in New Zealand. In terms of their relationship with, and regard for nature, Williams points out that the 'key to the Māori view towards environmental issues is the importance of not altering *mauri* [a resource's vital essence] to the extent that it is no longer recognizable' (2006, p. 74). Natural resource management, Williams suggests, can be fully attuned with human demands, but it must be achieved such that areas being harvested do not lose their essential character as a result of the harvest (2006). This worldview is somewhat incompatible with how New Zealand industries continue to intensify their land and water use, which often leads to a significant diminution of water quality. By-products of industry, most notably dairy farming, have progressively imperiled the integrity of rivers, lakes, and aquifers (Houlbrooke et al., 2004). Environmental degradation, as Baskaran et al. suggest, has also been coupled with increases in ground and surface water demands, as well as in acute reductions in pastoral landscapes in New Zealand (2009).

The principles that govern Māori views on natural resource management are enshrined in New Zealand's foundational document, the Treaty of Waitangi, where the Crown recognizes the authority that *iwi* (tribes) exert over their lands and natural resources (Ministry of Justice, 2016). Reconciling human needs with the long-term health of nature is mandatory in fulfilling a Māori worldview, but these customs are often incompatible with Pākehā views on legal ownership and resource management. These differences are most evident in the Stage 1 Report on the National Freshwater and Geothermal Resources Claim before the Waitangi Tribunal, which shows that western-style principles of resource ownership do not fully reflect the Māori tenets of authority, stewardship, and control over natural resources (Waitangi Tribunal, 2012). These incompatibilities, as Smith argues, do not emanate solely from misunderstandings between Māori and Pākehā groups: They derive largely from a European application of a 'moral perspective to land ownership and management, of 'use it or lose it'' (2012, p. 56) that inherently circumvents traditional Māori values that amalgamate ideas of culture and identity with natural resource management. On Māori's close connection with nature and how this connection relates to their identity, the New Zealand Māori Council writes:

> Land provides us with a sense of identity, belonging and continuity. It is proof of our continued existence not only as a people, but also as tangata whenua [original inhabitants] of this country. It is proof of our tribal and kin group ties. Māori land represents turangawaewae [place where one has the right to stand]. It is proof of our link with the ancestors of our past, and with generations to come. It is an assurance that we shall forever exist as a people, for as long as the land shall last. (1983, p. 10).

Considering the deep connections that Māori have with water resources, and nature in general, it is understandable how and why intra-state water conflicts between these groups could emerge in a jurisdictional setting like New Zealand's. If water access criteria are arrayed in favor of commercial enterprises that threaten the integrity of water resources as per Māori philosophy, both the well-being and identity of Māori water users become threatened, thus increasing the potential for conflict along Māori-Pākehā lines.

This, however, is not to say that water disputes in New Zealand are exclusive to the Māori-Pākehā polarity: Tensions over water use also exist within these groups. Such incompatibilities, in this case within Māori *iwi*, were most evident in 2013 when the South Island *iwi*, Ngāi Tahu, sought to invest in a water catchment project in the Makaroro River in Hawke's Bay, a venture that was heavily resisted by a local North Island *iwi*, Ngāti Kahungunu (Stuff, 2013). The project, conceived to intensify water and land use for agricultural purposes (Farmers Weekly, 2016), was perceived by Ngāti Kahungunu as detrimental to the quality of their ground and surface water. An agreement was later reached by both *iwi*, where Ngāi Tahu conceded to withdraw support from the project if the venture did not abide by the same environmental and cultural standards that Ngāi Tahu applied to their own rivers (Te Rūnanga o Ngāi Tahu, 2013). Even though the project was subsequently abandoned by Ngāi Tahu and its collaborating partners, this case highlights the extent to which inter-group incompatibilities over water can emerge within relatively like-minded groups.

Akin to grievances between Māori *iwi*, incompatibilities are also discernible between sectors seeking access to natural resources for financial gain. For example, New Zealand's National Institute of Water and Atmospheric Research (NIWA) notes that several conflicts have arisen in recent years over the development of the fish-farming industry in New Zealand, which has antagonized other aquaculture activities, most notably mussel farming (2013). Similarly, civil society groups concerned with resource intensification for animal-based agriculture assert that enterprises such as dairy farming have led to an exponential increase in water use. In their view, this usage jeopardizes water availability for other purposes, including plant-based food production (Vegans New Zealand, 2016).

The above elaboration leads to one central conclusion: Despite freshwater surplus and relatively strong institutional and procedural frameworks for water management, New Zealand is not immune to water-related conflicts. It is within such conditions that this study finds its theoretical and empirical anchors.

Water Commercialization in New Zealand

As per this chapter, the causal relationship between the commercialization of abundant natural resources and the emergence of conflict is of acute interest to abundance theorists. Applying this focus to freshwater in New Zealand requires the formulation of a question that assesses the dyadic relationship between freshwater commercialization and conflict. To examine such a question scientifically would require an acknowledgment that commercial enterprises vary significantly in their profit-generating capacities, water extraction and use, and exclusion potential. In this regard, commercial enterprises are bound to exert an asymmetrical effect on the emergence of water-related incompatibilities between individuals, groups, and sectors.

One cannot begin to discuss the water commercialization-conflict polarity in New Zealand without alluding to the development of the dairy industry, arguably one of the principal economic drivers in the country (New Zealand Institute of Economic Research, 2017; Statistics New Zealand, 2017). While significant focus has been dedicated to the dairy industry as a cause of environmental degradation (Davies-Colley & Nagels, 2002; Davies-Colley et al., 2004; Larned et al., 2004), one cannot convincingly argue that all dairy enterprises display the same conflict-inducing potential. A small dairy operation, to name one example, is unlikely to motivate skirmishes over its water access and use based on its low environmental and social impacts. As dairying activity intensifies its production, however, its demands for water increase (Allan, 2004). This generates concerns about water access, scarcity, and opportunity costs for other water consumers. For good or ill, New Zealand's dairy industry has displayed precisely this progressive intensification: Assessments of land use per industry demonstrate that between 1993 and 2012, land use for dairying in New Zealand increased by 46% (Statistics New Zealand in Foote et al., 2015), with the total amount of dairy cows reaching an estimated 6.5 million in 2012, up from 3.4 million in 1990 (Statistics New Zealand in Foote et al., 2015). This upward trend is consistent with the period covering 2012 and 2016, when land use for dairying increased by an additional 42.4% (Statistics New Zealand, 2018).

As the dairy industry intensifies, questions have emerged concerning the access and use parameters of other users and sectors dependent on freshwater. For example, New Zealand's Ministry for the Environment notes that several waterways near pastoral lands consistently contain unacceptable levels of E. coli, rendering those waterways unfit for activities such as swimming and whitebait fishing (2009a). Similarly, New Zealand's Parliamentary Commissioner for the Environment notes that fecal contamination can affect drinking water, quality of shellfish and can have negative impacts on the health of livestock (2004).

The direct and indirect effects of intensive agriculture (and other activities) in New Zealand have generated conversations over how to securitize the country's drinking water systems; these discussions have been rooted in the introduction of water treatment strategies such as chlorination. As district councils in New Zealand fail to comply with nationally mandated drinking water quality standards, frictions have emerged over both the setting of water quality guidelines and the costs attached to water treatment infrastructure and administration. This was the case in the Thames-Coromandel district, where local authorities argued that the central government's demands for higher water quality standards would inevitably bear additional costs for local taxpayers. To ensure compliance, the Thames-Coromandel District Council included a scheme to chlorinate all water supplies in its 2018–2028 plan, including the last two unchlorinated sites: the towns of Hahei and Pauanui (Newsie, 2017). The latter town's water supplies were subsequently chlorinated through the provision of equipment by Southern Cross Consulting Ltd., a company that specializes in water chlorination for municipal and community-level use (Southern Cross Consulting Ltd, 2017). This example illustrates two important points. Firstly, water authorities in New Zealand are becoming incapable of guaranteeing the good quality of the water under their jurisdiction, and secondly, it showcases the role of the private sector in the provision and administration of water security strategies. In short, the commercialization of freshwater in New Zealand encompasses several dimensions of water resources, from their industry-driven degradation, to the fashion in which city/district and regional councils have chosen to securitize it for human use. It is in this context that this study argues conflicts may emerge between individuals and groups affected by these activities.

Competing Values and Interests

When examining the question of water conflicts driven by enterprises such as dairy farming or activities such as water treatment strategies, one is ultimately assessing the frictions and tensions that emanate from individuals and groups that espouse competing interests over water access and use. This relates tacitly to Peters' argument that resource-driven conflicts are extrapolations of competing meanings systems. On this note, Peters observes the following:

> Claims to use and control resources and to exercise authority over things and people are premised on an ideology or a set of meanings. Struggles over resources or over power, then, necessarily take place in terms of such meanings. These meanings are both shared and disputed: different categories or groups assign different meanings, different definitions or different emphases at different times to known concepts, events and acts. Hence, one event, one institution or one concept may be defined and interpreted in a number of ways and in ways that contradict each other. (1984, p. 29).

The values or meanings to which Peters alludes determine how and why a specific group seeks to interact with nature and its diverse elements. Individuals and groups (in New Zealand and elsewhere) are drawn to water for spiritual (Kamitsis & Francis, 2013), cultural (Jahren & Sui, 2017), and religious reasons (Chamberlain, 2008), while others may identify water more instrumentally in terms of food requirements (Gonzalez-Dugo et al., 2010), sanitation (WHO, 2017), and industrial use (USGS, 2014). Household-level demands are similarly reliant on adequate access to water resources (Mayer et al., 1999; EPA, 2017). Other groups are driven toward the small-scale use of natural resources and, in some instances, on their non-use or conservation.

For example, Ravindra et al. describe the people of Rajasthan (India) as a group that assigns value to water based on environmental conservation and community building, rather than on commercial yield. These values are reflected in infrastructural developments, known as Johads, that are conceived to replenish aquifers via rainwater (Ravindra et al., 2017) and to ensure that local communities are central in the equitable and fair distribution of the preserved water (Hussain et al., 2014). Similar water catchment systems known as 'Aflaj' are employed in Oman (Al-Marshudi, 2007), with priority of use given to direct consumption, sanitation, proper disposal of the dead, and food production, and at no cost for the system's beneficiaries (Trapasso, 1996). In these systems, profit-making is not the goal of water access, and the infrastructure that facilitates said access is equally defined by non-commercial criteria.

As in Rajasthan and Oman, similar dynamics are discernible in New Zealand, where pressures to prioritise ecological health and cultural preservation have led to a number of policies. One example is the 2017 passing of the Te Awa Tupua Bill, which conferred legal personhood to the Whanganui River and put measures in place to ensure its restoration and protection (New Zealand Parliament, 2017). A similar arrangement was later made through the Record of Understanding for Mount Taranaki, Pouakai, and the Kaitake Ranges, where legal personhood was given to Mount Taranaki and its surrounding areas (Nga Iwi and The Crown, 2017). This spirit is also reflected in recent changes to New Zealand's freshwater management system. Of relevance are amendments to the National Policy Statement for Freshwater Management, which introduce the concept of *te mana o te wai*. *Te mana o te wai* provides a framework in which decisions over how freshwater is accessed and used must prioritize the health and well-being of waterbodies prior to its other uses (Ministry for the Environment, 2020). This is premised on the belief that if waterways are cared for, all other uses and needs will be implicitly satisfied. These examples bear witness to the types of initiatives that prioritize conservation and care above other utilitarian values.

As per the previous examples, the multitude of values that can be attached to water resources may be as varied as the individuals and groups who seek access to them. This heterogeneity does not seem to be reflected in the way environmental security analysts approach the study of natural resource abundance as a conflict-inducing factor. The previous chapter emphasized the literature's implicit focus on non-renewable resource commercialization as a causal determinant of conflict. However, when one considers freshwater abundance or, more specifically, the commercialization of abundant freshwater resources, there are other factors that dictate the rise of conflict. Indeed, these factors encompass the economic dimension of freshwater but, unlike non-renewables, are not fully defined by it. This is a reality that arguably circumscribes water issues, both in New Zealand and elsewhere.

Non-renewables, such as minerals and fossil fuels, are crucial for the sustainment of modern human livelihoods, and these demands partially explain why such resources are generally coupled with a high economic value. For example, the US Energy Information Administration (EIA) predicts that by 2040, the global demand for energy will rise by 48%, with liquid fuels, natural gas, and coal accounting for 78% of this increase (2016). In New Zealand, reliance on non-renewables follows a similar upward trend, with the consumption of coal, diesel, premium petrol, and natural gas seeing upsurges in the years leading up to 2016 (New Zealand Ministry of Business, Innovation and Employment, 2017). In a world where supply and demand dictate global rhythms of resource exploitation and where societies have not yet fully adopted a post-fossil fuel agenda, it is understandable why environmental security literature devotes such time and space to the conflict-inducing potential of non-renewables. However, applying this approach to freshwater is problematic because emphasizing water's economic character often leads to a negation of other values typically assigned to water.

As explained above, an abundance of freshwater resources poses the plausibility of a theoretical scenario of sufficiency for all individuals and groups. However, the way several commercial enterprises are seizing New Zealand's freshwater resources challenges the sufficiency argument. Unsatisfied water demands in a resource-rich environment are largely attributed to the prioritization of commercial interests, rather than water scarcity. Water abundance, on the other hand, focuses the attention of user groups on water quality and quantity and how these connect to industrial and commercial growth patterns. In such a setting, manufactured rather than natural inadequacies appear to drive actors into conflicts over water.

To explore how conflict emerges from water commercialization, this research proposes the following research question: How and why does the commercialization of freshwater affect the emergence of hydropolitical conflict in New Zealand?

References

Al-Marshudi, A. S. (2007). The falaj irrigation system and water allocation markets in Northern Oman. *Agricultural Water Management, 91*, 71–77.

Allan, J. D. (2004). Landscapes and riverscapes: The influence of land use on stream ecosystems. *Annual Review of Ecology, Evolution, and Systematics, 35*, 257–284.

Ball, A. (2007). *Estimation of the burden of water-borne disease in New Zealand: Preliminary report*. Wellington. Retrieved March 8, 2022 from https://www.health.govt.nz/system/files/documents/publications/water-borne-disease-burden-prelim-report-feb07-v2.pdf

Baskaran, R., Cullen, R., & Colombo, S. (2009). Estimating values of environmental impacts of dairy farming in New Zealand. *New Zealand Journal of Agricultural Research, 52*(4), 377–389.

Bermejo, R. (2014). *Handbook for a sustainable economy*. Springer.

Bond, P., & Dugard, J. (2008). Water, human rights and social conflict: South African experiences. *Law, Social Justice & Global Development Journal*, (1).

Bourne, P. G. (1984). Water and sanitation for all. In P. G. Bourne (Ed.), *Water and sanitation: Economic and sociological perspectives*. Academic Press Inc.

Chamberlain, G. L. (2008). *Troubled waters: Religion, ethics, and the global water crisis*. Rowman & Littlefield.

Collins, D., Horrell, G., & Pearson, C. (2015). *Surface water components of New Zealand's national water accounts, 1995–2014*. Christchurch.

Davies-Colley, R., & Nagels, J. (2002). Effects of dairying on water quality of lowland stream in Westland and Waikato. In *Proceedings of the New Zealand grasslands association* (pp. 107–114).

Davies-Colley, R. J., Nagels, J. W., Smith, R. A., Young, R. G., & Phillips, C. J. (2004). Water quality impact of a dairy cow herd crossing a stream. *New Zealand Journal of Marine and Freshwater Research, 38*(4), 569–576.

Department of Internal Affairs. (2021). *Three waters review.* Retrieved January 24, 2022, from https://www.dia.govt.nz/Three-waters-review

EIA. (2016). *International energy outlook 2016.* Washington DC. Retrieved August 9, 2018, from https://www.eia.gov/outlooks/ieo/pdf/0484(2016).pdf

EPA. (2017). *How we use water.* Retrieved May 25, 2017, from https://www.epa.gov/watersense/how-we-use-water#DailyLife

Farmers Weekly. (2016). *Preferred investor "close" for Ruataniwha dam.* Retrieved November 20, 2017, from https://farmersweekly.co.nz/#

Foote, K. J., Joy, M. K., & Death, R. G. (2015). New Zealand dairy farming: Milking our environment for all its worth. *Environmental Management, 56*(3), 709–720.

Gluckman, P. (2017). *New Zealand's fresh waters: Values, state, trends and human impacts.* Auckland. Retrieved March 23, 2018, from http://www.pmcsa.org.nz/wp-content/uploads/PMCSA-Freshwater-Report.pdf

Gonzalez-Dugo, V., Durand, J.-L., & Gastal, F. (2010). Water deficit and nitrogen nutrition of crops. A Review. *Agronomy for Sustainable Development, 30*, 529–544.

Gustafsson, J.-E. (2001). *Public water utilities and privatisation in Sweden. EPSU public service conference.* Brussels.

Houlbrooke, D. J., Horne, D. J., Hedley, M. J., Hanly, J. A., & Snow, V. O. (2004). A review of literature on the land treatment of farm-dairy effluent in New Zealand and its impact on water quality. *New Zealand Journal of Agricultural Research, 47*, 499–511.

Hussain, J., Husain, I., & Arif, M. (2014). Water resources management: traditional technology and communities as part of the solution. *In Proceedings of the International Association of Hydrological Sciences, 364*, 236–242.

Jahren, P., & Sui, T. (2017). Water and Culture. In P. Jahren & T. Sui (Eds.), *How Water Influences our Lives.* Singapore: Springer.

Kamitsis, I., & Francis, A. J. (2013). Spirituality mediates the relationship between engagement with nature and psychological wellbeing. *Journal of Environmental Psychology, 36*, 136–143.

Larned, S. T., Scarsbrook, M. R., Snelder, T. H., Norton, N. J., & Biggs, B. J. F. (2004). Water quality in low-elevation streams and rivers of New Zealand: Recent state and trends in contrasting land-cover classes. *New Zealand Journal of Marine and Freshwater Research, 38*, 347–366.

Lawrence, P., Meigh, J., & Sullivan, C. (2002). Keele economics research papers the water poverty index: An international comparison. *Keele Economics Research Papers, 19.* Retrieved June 1, 2017, from http://econwpa.repec.org/eps/dev/papers/0211/0211003.pdf

Mayer, P. W., DeOreo, W. B., Opitz, Eva, M., Kiefler, J. C., Davis, W. Y., Dziegielewski, B., & Nelson, J. O. (1999). *Residential end uses of water*. Denver.

Ministry for the Environment. (2009a). *Water quality in selected dairy farming catchments a baseline to support future water-quality trend assessments Acknowledgements.* Wellington. Retrieved June 11, 2018, from https://environment.govt.nz/publications/water-quality-in-selected-dairy-farming-catchments-a-baseline-to-support-future-water-quality-trend-assessments/

Ministry for the Environment. (2009b). Your guide to the environment court. *An everyday guide to the RMA.* Retrieved March 13, 2022, from https://www.environmentcourt.govt.nz/assets/Documents/Publications/MFE-Your-Guide-to-the-Environment-Court.pdf

Ministry for the Environment. (2017). *New Zealand's environmental reporting series: Our fresh water 2017.* Retrieved June 20, 2017, from https://environment.govt.nz/publications/our-fresh-water-2017/

Ministry for the Environment. (2020). *National policy statement for freshwater management 2020.* Wellington. Retrieved February 5, 2022, from https://environment.govt.nz/assets/Publications/Files/national-policy-statement-for-freshwater-management-2020.pdf

Ministry for the Environment. (2022a). *National direction.* Retrieved February 5, 2022a, from https://environment.govt.nz/acts-and-regulations/national-direction/

Ministry for the Environment. (2022b). *Kia kaha ake te tiakina o ngā puna wai-inu / Improving the protection of drinking-water sources: Proposed amendments to the Resource Management (National Environmental Standards for Sources of Human Drinking Water) Regulations 2007: Consultation document*. Wellington. Retrieved February 5, 2022b, from https://consult.environment.govt.nz/freshwater/nes-drinking-water/

Ministry of Business Innovation & Employment. (2017). *Energy in New Zealand 2017*. Wellington. Retrieved October 30, 2017, from http://www.mbie.govt.nz/info-services/sectors-industries/energy/energy-data-modelling/publications/energy-in-new-zealand/documents-images/energy-in-nz-2017.pdf

Ministry of Health. (2021). *Drinking-water*. Retrieved March 12, 2022, from https://www.health.govt.nz/our-work/environmental-health/drinking-water

Ministry of Justice. (2016). *Meaning of the treaty*. Retrieved October 26, 2017, from https://www.waitangitribunal.govt.nz/treaty-of-waitangi/meaning-of-the-treaty/

New Zealand Institute of Economic Research. (2017). *Dairy trade's economic contribution to New Zealand*. Wellington. Retrieved October 27, 2017, from https://nzier.org.nz/static/media/filer_public/29/33/29336237-3350-40ce-9933-a5a59d25bd31/dairy_economic_contribution_update_final_21_february_2017.pdf

New Zealand Māori Council. (1983). *Kaupapa: Te Wahanga Tuatahi*. New Zealand Maori Council.

New Zealand Parliament. (2017). Innovative bill protects Whanganui River with legal personhood. Retrieved May 9, 2019, from https://www.parliament.nz/en/get-involved/features/innovative-bill-protects-whanganui-river-with-legal-personhood/

Newshub. (2016). *Hundreds protest against Ashburton water sale*. Retrieved February 23, 2018, from http://www.newshub.co.nz/home/new-zealand/2016/04/hundreds-protest-against-ashburton-water-sale.html

Newsie. (2017). *Coromandel water supplies in question*. Retrieved April 15, 2019, from https://www.newsie.co.nz/news/71083-coromandel-water-supplies-question.html

NewsTalkZB. (2017). *Environment Canterbury responds to water bottling petition*. Retrieved January 9, 2018, from http://www.newstalkzb.co.nz/news/christchurch/environment-canterbury-responds-to-water-bottling-petition/

Nga Iwi, & The Crown. (2017). *Record of Understanding for Mount Taranaki, Pouakai and the Kaitake Ranges*. Retrieved May 5, 2019, from https://www.govt.nz/dmsdocument/7265-taranaki-maunga-te-anga-putakerongo-record-of-understanding-20-december-2017

NIWA. (2013). *Down on the Farm*. Retrieved June 30, 2018, from https://www.niwa.co.nz/publications/wa/water-atmosphere-7-june-2013/down-on-the-farm

Parliamentary Commissioner for the Environment. (2004). *Growing for good: Intensive farming, sustainability and New Zealand's environment*. Wellington. Retrieved June 11, 2018, from http://www.pce.parliament.nz/media/pdfs/growing-for-good.pdf

Parliamentary Counsel Office. (1991). *Resource Management Act 1991*. Retrieved March 13, 2022, from https://www.legislation.govt.nz/act/public/1991/0069/latest/DLM2416413.html?search=sw_096be8ed81baa4cf_affected+person_25_se&p=1

Peters, P. E. (1984). Struggles over water, struggles over meaning: Cattle, water and the state in Botswana. *Africa, 54*(3), 29–49.

Ravindra, A., Raju, K. V., & Manasi, S. (2017). Policy perspectives: Water and scriptures. In K. V. Raju & S. Manasi (Eds.), *Water and scriptures* (pp. 247–267). Springer International Publishing.

Smith, P. M. (2012). *A concise history of New Zealand* (2nd ed.). Cambridge University Press.

Southern Cross Consulting Ltd. (2017). Clean, safe, affordable. Retrieved April 15, 2019, from https://chlorinationnz.com/

Statistics New Zealand. (2005). *Population. New Zealand in the OECD*. Wellington.

Statistics New Zealand. (2011). *Water physical stock account: 1995–2010*. Wellington.

Statistics New Zealand. (2017). *Global New Zealand—International trade, investment, and travel profile: Year ended December 2016*. Wellington.

Statistics New Zealand. (2018). *Agricultural and horticultural land use*. Retrieved June 12, 2018, from http://archive.stats.govt.nz/browse_for_stats/environment/environmental-reporting-series/environmental-indicators/Home/Land/land-use.aspx

Stuff. (2013). *Ngai Tahu follows the water*. Retrieved May 14, 2019, from http://www.stuff.co.nz/business/industries/9200166/Ngai-Tahu-follows-the-water

Stuff. (2018). *The South Island towns (and city) fighting chlorinated drinking water*. Retrieved May 21, 2019, from https://www.stuff.co.nz/national/health/104005202/the-south-island-towns-and-city-fighting-chlorinated-drinking-water

Tanentzap, A. J., Lamb, A., Walker, S., & Farmer, A. (2015). Resolving conflicts between agriculture and the natural environment. *PLOS Biology, 13*(9), 1–13.

Te Rūnanga o Ngāi Tahu. (2013). *Joint release between Ngāti Kahungunu, Te Rūnanga o Ngāi Tahu and Ngāi Tahu Holdings Corporation*. Retrieved May 14, 2019, from https://ngaitahu.iwi.nz/joint-release-ngati-kahungunu-te-runanga-o-ngai-tahu-ngai-tahu-holdings-corporation/

The Environment Court of New Zealand. (2022). *Overview of the environment court's processes*. Retrieved February 13, 2022, from https://environmentcourt.govt.nz/court-process/

Trapasso, L. M. (1996). Water wisdom of the ancients. *Civil Engineering, 66*(1), 64–65.

USGS. (2014). *Estimated use of water in the United States in 2010*. Reston.

Vegans New Zealand. (2016). *Dairy is damaging New Zealand*. Retrieved November 21, 2017, from http://www.vegans.co.nz/vegan-blog/dairy-is-damaging-new-zealand/

Waitangi Tribunal. (2012). *The report on the national freshwater and geothermal resources claim*. Lower Hutt. Retrieved October 26, 2017, from https://forms.justice.govt.nz/search/Documents/WT/wt_DOC_59941926/Wai2358W.pdf

WHO. (2017). *WHO—Water and sanitation*. Retrieved July 24, 2017, from http://www.who.int/ceh/risks/cehwater2/en/

Williams, J. (2006). Resource management and Maori attitudes to water in southern New Zealand. *New Zealand Geographer, 62*, 73–80.

Chapter 3
Research Design

Abstract This chapter provides an overview of the research design through which the proposed arguments are examined. The design includes, first, a conflict intentionality and engagement analysis, where participants' perceptions over water bottling and water chlorination are assessed against their willingness to engage in conflict over those activities and at what level of intensity. Second, it employs a comparative analysis to identify common causal themes across the above cases. And third, this study applies a conflict intentionality classification that categorizes actors in relation to the level and type of hydropolitical conflict in which they are willing to engage or in which they became involved.

Introduction

The previous chapter explained the challenges faced by individuals and groups in New Zealand over the management and use of the country's abundant water supplies. The proliferation and growth of enterprises such as water bottling and industrial dairy farming raises concerns around the quantitative and qualitative state of New Zealand's freshwater. As such, the Coleridge Dilemma as described in this chapter seems to apply in the immediate New Zealand context: There is a misalignment between freshwater surplus, its management, and the extent to which water demands are satisfied. Examination of these issues as they apply to New Zealand is therefore necessary to determine the drivers of water-related disputes.

To explain the research approach employed in this book, this chapter will present the framework through which water disputes in New Zealand will be analyzed. The framework draws heavily, at least from a structural standpoint, from King, Keohane, and Verba's proposed design components, which include the research question, theory, data, and use of the data (1994). These components are elaborated in no specific order or hierarchy throughout this chapter.

A. E. Suazo, *The Risk of Water Conflicts in Aotearoa-New Zealand*,
https://doi.org/10.1007/978-3-031-06660-3_3

Research Question

In light of the observations elaborated in the previous chapter, this study formulates the following research question: *How and why does the commercialization of freshwater affect the emergence of hydropolitical conflicts?* In the context of this book, water commercialization is defined as an activity or enterprise that derives direct or indirect monetary benefit from the exploitation and/or manipulation of freshwater resources. This definition is proposed to enable the analysis of a wide spectrum of enterprises, ranging from activities that seek access to water for its direct marketization (e.g., water bottling, bulk water sale, soft drink production), that draw indirect revenue from water use (e.g., plant and animal-based food production), and that manipulate water in a way that, directly or indirectly, provides financial benefits for some parties (e.g., externalization of production costs via water pollution, water chlorination, water fluoridation).

The identification of water disputes will be enabled through the application of Eidem, Fesler, and Wolf's definition of hydropolitical conflict, which they describe as 'any [negative] interaction between parties that is action-defined, recorded, and made available to the public [that is] driven by some aspect or dimension of fresh water resources' (Eidem et al., 2012, p. 63). To complement their definition, Eidem, Fesler, and Wolf propose a hydropolitical conflict spectrum that categorizes events in relation to their intensity. They refer to this spectrum as the Conflict-Cooperation Intensity Scale (from this point on referred to as the CCI Scale), a tabulation of which is presented below (Table 3.1).

Eidem, Fesler, and Wolf's definition of hydropolitical conflicts, and its adjoined CCI Scale, proves beneficial because they facilitate the examination of a wide array of water-related conflict events, spanning from low-scale manifestations of discontent and disagreement, to overt episodes of violence. By applying this definition, one can expand the study's analytical space beyond the literature's traditional focus on violent forms of conflict. The definition is also expected to enable a thorough investigation of other less-identifiable forms of water-driven, non-violent discord. Together, these tools also provide an opportunity to account for hydropolitical conflict intensity in a dynamic fashion, allowing the analysis to identify the escalation of conflict dynamics as they unfold.

In the immediate New Zealand context, it is important to note that litigations and appeals, the second-most intense types of conflict featured in the CCI Scale, are processes sanctioned by the RMA. Conflicts at this level escalate beyond the immediate purview of local authorities and become administered by central judicial processes and institutions, including New Zealand's Environment Court and High Court.

Table 3.1 Conflict-cooperation intensity scale

Intensity level	Event type
−5	Small scale acts of violence, protests, vandalism
−4	Litigations, appeals
−3	Fines, proposal and permit denials, halting negotiations
−2	Petitions, withdrawal of third-party support
−1	Delays, report reviews, voicing opposition, editorials
0	Judicial rulings, no comment statements, announcements
1	Voicing opinion of approval, court-forced negotiation, editorial
2	Meetings, third-party-support, negotiation requests
3	Permit approvals, fixing violations, negotiations begin
4	Lawsuit settlements, regulation approval, management transfers
5	State bill passage, compacts or official agreements

The Value of Qualitative Research

Chapter 1 explained the literature's overemphasis on quantitative methodologies to examine the natural resource abundance-conflict polarity. This was argued to be the case due to the literature's asymmetrical focus on abundant non-renewable resources as causal determinants of conflict and on the financial value systems normally attached to these types of resources.

In their own right and in their own contexts, these methodologies have brought tremendous value to Abundance Theory. However, it is considered that quantitative methodologies are ill-suited to examine the proposed research question, because it risks downplaying the non-commercial values that influence individuals' and groups' drive to access and use freshwater and how and why these values may conflict with one another.

Quantitative research poses an additional challenge in the immediate New Zealand context; the Treaty of Waitangi confers special authority upon Māori populations with regards to the management and guardianship of natural resources, including freshwater (Ministry of Justice, 2016). While some commercial interests indeed influence the access and use criteria of some Māori *iwi* (Stuff, 2013), Māori communities also assign different non-commercial values to their freshwater, values that may not be adequately captured through quantitative instruments. Indeed, Māori scholars have highlighted the tensions involving the commercialization of freshwater and its impact on Māori communities. For example, Muru-Lanning explains that industrial development along the Waikato River has led to Māori people becoming alienated from it (2012). Similar views are expressed by Poata-Smith when arguing that Māori are likely to become alienated from their local waterways because of the increased bureaucratization of water bodies by local authorities (2004). The question of alienation, as a source of Māori concern, is a prime example of the types of incompatibilities over water access and use that are difficult to capture and examine through quantitative methodologies.

Therefore, applying a quantitative approach to the study of freshwater and conflict in New Zealand is likely to produce an analysis that disproportionately emphasizes the commercial character of freshwater, at the expense of other non-commercial value systems central to Māori people. Employing quantitative tools is expected to produce observations that inaccurately reflect Māori communities' views and interests and those of other individuals and groups that may see in Māori ideals a better alignment for their own value systems.

Considering the above observations, qualitative research was deemed the best methodological framework for the study of individuals' water access and use criteria. As such, qualitative research provides the necessary tools to relate and analyze the participants' views over how and why values and activities of different types may conflict with one another. It is considered that qualitative research will help uncover the multidimensionality inherent in water-related disputes, enabling the identification and analysis of the complexities, interdependencies, and ambiguities expected in the proposed inquiry. In this regard, Strauss argues that qualitative research is necessary to conduct a 'detailed, intensive, microscopic examination of the data to bring out the amazing complexity of what lies in, behind, and beyond those data' (Strauss in Aspers & Corte, 2019, p.148).

The use of qualitative research in the environmental conflict space is in no way novel. For example, Griewald and Rauschmayer employ a qualitative research design to investigate the causal links between deforestation and the emergence of conflict in Germany, which allowed them to capture and analyze different aspects of the participants' views regarding forest clearing (2014). Similarly, qualitative research has been deployed by researchers to examine individuals' perceptions and personal views over practices such as water bottling. For example, through the use of qualitative methodologies, Ragusa and Crampton find evidence that suggests that individuals from Australia and New Zealand have little trust in companies that bottle and sell water (2016). It is therefore anticipated that a qualitative framework will add similar value to the study of freshwater and conflict.

The Arguments

This study proposes two central arguments. First, that the introduction of water commercialization practices influences the emergence of hydropolitical conflict intentionality if the practice is economically, socially, and environmentally incompatible with the values of residents. In line with the explanations presented in the previous chapter, the proposed argument acknowledges that water commercialization practices differ tremendously in profit generation, scale of extraction/use, and exclusion potential and further suggests that some practices are more compatible with the economic, social, and environmental realities of communities.

Second, this study argues that hydropolitical conflict risk intensifies in accordance with the extent to which residents trust local approval and appeals processes behind water commercialization practices. This study understands risk intensification as the displacement of a conflict event or intentionality from one level of the CCI Scale to at least one of greater severity. To test this argument, this study will classify hydropolitical conflict events in three categories: high-intensity conflicts as events that fall between the −4 and −5 categories of the CCI Scale, medium-intensity events will be included in the −3 category, and low-intensity conflicts will fall between the −1 and −2 levels of the scale. No comment and no opinion statements will be assigned to the 0 category of the CCI Scale.

This study further defines an approval and appeals process as the set of locally controlled legislative and procedural mechanisms through which a water-based enterprise becomes realized and that permits citizens to voice support or disapproval for said enterprise. In this regard, this study anticipates that commercial operations approved under clearly delineated and trust-inspiring approval and appeals processes are less likely to motivate the escalation of risk to the top-most levels of the CCI Scale, than operations that are approved under dubious decision-making processes.

Operationalization and Case Selection Strategies

To investigate the causal effect of water commercialization on the emergence and intensification of water conflicts, this study operationalizes water commercialization by examining the introduction of water bottling and water chlorination in the towns of Ashburton and Glenorchy. These enterprises will be explored to elucidate the relationship between water commercialization and conflict and water security and conflict. These activities also stand out because they display a considerable variation in terms of profit generation, scale, and the extent to which they affect local communities' water access. Thus, it is expected that the two dynamics will have an escalatory impact on hydropolitical conflict risk.

Because the proposed research questions examine commercial enterprises, this study will only investigate the impact of operations that draw a financial gain from access to or manipulation of, freshwater. As such, it locates water bottling and water chlorination at opposite sides of the for-profit spectrum. For instance, water bottling is considered an enterprise that derives significant rent from high water extraction. It is also proposed that water bottling is highly exclusive because water access is dependent on a monetary exchange and imposes significant limitations on its use.

Similarly, this study regards water chlorination as a commercial activity because drivers have been identified that relate public water treatment strategies to private commercial gains. In the context of this study, and of New Zealand, the act of chlorinating a municipality's water supplies is often carried out under a public–private stakeholder arrangement, where public water quality standards are at least partly facilitated by private enterprises specializing in the administration of chlorine and the provision of equipment and maintenance. A good example of such an enterprise is Apex Environmental, a company that describes itself as being 'in the business of designing, building, installing and commissioning water and wastewater treatment plants for the food and beverage, dairy, textiles, winery, and municipal sectors' (Apex Environmental, 2017, p. 3). Given this and the continued pressures posed by water sector representative groups (Water New Zealand, 2018) and government entities (Department of Internal Affairs, 2017; Harris, 2017) to implement universal water treatment schemes in New Zealand, one can anticipate that the increasing trend to securitize water resources through chlorination is likely to be a driver for a greater participation from, and commercial gains for, the private sector.

Like water bottling, whose scale is defined by its level of water extraction, water chlorination is considered to have a similarly high scope, due to the wide-ranging social impacts on a locality's water supplies. Whereas the rent and scale characteristics of chlorination bear similar values to those of water bottling, both enterprises differ in their exclusion potential. Unlike water bottling, water chlorination is designed as a measure to make a specific location's public water supplies safer and is not conceived to alter pre-existing access or procurement criteria. Despite contemporary research that stresses the different types of health risks linked to chlorinated water (Abbas et al., 2015; Legay et al., 2011; Rifkin & Bouwer, 2007), the act of chlorinating public water supplies is not a policy conceived to curtail water access and therefore is not regarded by this study as bearing the same exclusion potential inherent in a water bottling operation.

The introduction of water chlorination in a water-abundant environment offers unique insights into how the Coleridge Dilemma unfolds in a country such as New Zealand, where water-related concerns are far-removed from questions of water quantity. Rather, they are more closely related to concerns over water quality degradation caused by the activities of individuals, groups, and sectors. As such, water chlorination is not a proxy of water abundance, but it provides a useful entry point to analyze the perceived sources of water quality decay, the measures that authorities enact to improve water quality in their jurisdictions, and the conditions under which these measures can influence the emergence and intensification of hydropolitical conflicts.

Case Studies

This study will investigate the causal effects of water bottling and water chlorination in the towns of Ashburton (Canterbury) and Glenorchy (Otago). These cases were chosen because both locations have introduced water bottling and chlorination processes at different stages. Within the Glenorchy township, for example, a company by the name of Koha Water Limited is seeking approval for a water bottling operation, which proposes the extraction and processing of roughly 236 million liters of artesian water per year until 2038 (Radio New Zealand, 2017), a process involving the approval of the Otago Regional Council (ORC). Additionally, the town's district-level water authority, the Queenstown-Lakes District Council (QLDC), initiated a water chlorination scheme to give effect to the recommendations by the Havelock North Drinking Water Inquiry Report (referred henceforth as Havelock North Report), which, among other guidelines, suggests that water authorities in New Zealand introduce universal treatment of their water supplies as a measure to prevent water-borne diseases (Department of Internal Affairs, 2017). The decision to chlorinate Glenorchy's public water supply was approved in December 2017 and implemented in January 2018.

As in Glenorchy, water bottling and chlorination have been envisaged and instituted for Ashburton at various points in time. In 2016, the water bottling firm NZ Pure Blue initiated plans to purchase part of a business estate owned and operated by the Ashburton District Council (ADC), with the intention of abstracting a calculated 1.4 billion liters of artesian water per year until 2046 (Stuff, 2016). The permit to extract the above volume was granted by the Canterbury Regional Council, also referred to by its trading name, Environment Canterbury (ECan) (Environment Canterbury, 2017). After the bottling plans were made public in April 2016, significant political mobilization took place in Ashburton, in some cases drawing crowds by the hundreds (Newshub, 2016). In addition to these mobilizations, Ashburton's residents organized a petition drive to help make the case against the bottling operation, an effort that succeeded in collecting an estimated 40,000 signatures (Stuff, 2016). Alongside water bottling, water chlorination in Ashburton was part of a broader effort by the ADC, dating back to 2003, to upgrade the water infrastructure under its jurisdiction. As a result, all 12 of the ADC's water supplies have been chlorinated on a permanent basis (Ashburton District Council, 2018).

These cases also prove beneficial for this study because they display significant variations in terms of their dependence on resource-intensive industries. Whereas Glenorchy, a town with a population of 363 (Statistics New Zealand, 2018a), features predominantly low-scale economic enterprises and various types of subsistence activities, Ashburton, with a population of roughly 13,500 (Statistics New Zealand, 2018b), lies in the heart of New Zealand's dairy and irrigation industries, both of which are heavily water-extractive. For example, data from Statistics New Zealand determines that an estimated 47.22% of Glenorchy's residents are employed in the accommodation and food services industry, while an additional 41.67% are employed in the arts and recreation services domain, narrowing employment in resource-dependent industries such as agriculture, forestry, and fisheries to 3.33% (Statistics New Zealand, 2018a). In contrast, the agricultural sector is the main driver of employment in Ashburton, with an estimated 21.5% of the population drawing some type of income from that industry (Statistics New Zealand, 2018b). By examining the effects of water bottling and water chlorination in Ashburton and Glenorchy, this study will be able to capture dynamics related to asymmetries in income generation, employment dependence relative to sector, and the type of historical and contemporary water use to which locals are accustomed in each town, all of which may play a part in the generation conflict intentionality and intensification of conflict risk in each location.

Methods of Analysis

To test the proposed arguments, this study applies three analytical instruments. First, it employs a conflict intentionality and engagement analysis to examine how the introduction of water bottling and water chlorination influences the respondents' willingness to engage in water-based conflicts and at which level of intensity. To do so, a 30-question guide (Appendix) was developed to capture the participants' general views of freshwater in their district, their attitudes toward water bottling and water chlorination, and their willingness to engage politically against those activities. When either enterprise inspired negative attitudes by the respondents, these attitudes were then juxtaposed with the participants' willingness to engage in hydropolitical conflict over that activity and what type of potential political action their engagement would entail. The study then asked participants to explain their perceptions over the decision-making process behind each activity. Finally, these accounts were distributed in relation to the categories of the CCI Scale, to determine both the levels of intensity of each participant's conflict intentionality, and once quantified, to get a sense of the community-wide state of public opinion regarding each operation.

To support the findings gathered through the question guide, this study also employed the analysis of a number of documents to ascertain the participants' actualized hydropolitical conflict engagement in both towns. These documents were selected based on the extent to which they reflected the participants' type of potential political engagement against water bottling and/or water chlorination. The chosen documents also provided insights into the level of scrutiny and transparency with which water authorities approved the examined practices. This in turn is expected to shed some light on the character of the approval and appeals process behind each water-based practice.

In view of the above requirements, this study examined the following documents:

- District and regional-level council documents and reports in both locations.
- Public meeting minutes.
- Official submissions documents.
- Newspaper articles.
- Internal business materials.

Second, this project employs a comparative analysis framework to examine how the causal impact of water bottling and water chlorination on conflict varies across cases. To this end, this study compares the participants' conflict intentionalities and examines any potential variations as perceived in Ashburton and Glenorchy.

A comparative framework permits a thorough investigation of the common threads found in each case and what causal outcomes may be empirically linked to them. More broadly, a comparative framework was chosen to allow for theoretical generalizations that are sensitive to the local and regional specificities of each location of study.

The value of comparative research in environmental politics has been highlighted previously by scholars such as Steinberg and VanDeveer. On the value of comparative research for the study of environmental phenomena, they argue that '[c]omparative research, at its best, occupies [a] position between theoretical generalization and an appreciation for the importance of context' (Steinberg & VanDeveer, 2012, p. 9) and that 'it relates particular empirical instances to broader theories by making systematic comparisons across political units' (Steinberg & VanDeveer, 2012, p. 9). By comparing the effects of water bottling and water chlorination in Ashburton and Glenorchy, it will be possible to capture local and regional specificities related to water access and use and to apply its results in locations featuring similar conditions.

And third, this study establishes a conflict intentionality classification to display the different types of actors that emerge in response to the introduction of a water bottling and water chlorination. To do so, this study divides the negative (conflict) side of the CCI Scale into events of low, medium, and high intensity. It then assigns the participants to each category based on the type and intensity of their conflict intentionality. Once the participants are located in each category, commonalities between the actors are assessed qualitatively to identify shared traits, views, objectives, and other properties that may become salient. In this analysis, accounts that report political engagement in the positive (cooperation) side of the CCI Scale will also be examined.

Together, it is expected that these tools will provide insights into how conflict intentionality over water bottling and chlorination emerges and how hydropolitical conflict risk may intensify as a result of these activities.

Data Collection

Because the proposed arguments focus on residents' political responses motivated by water bottling and water chlorination, this study tried to capture the voices of as many residents as possible in each location, to gain the best possible understanding of how and why hydropolitical conflicts emerge and intensify. In this instance, this study did not endeavor to specifically capture the voices of, for example, individuals that would consider themselves political activists, nor those of individuals employed in any one industry. Rather, its efforts were devoted to capturing the perspectives of as wide a variety of individuals as possible.

To achieve this research goal, a recruitment strategy was developed, where local community members were approached in casual settings, which included parks, farms, public libraries, art galleries, museums, and local businesses. These approaches were made at random, without prior knowledge of the individuals' political alliances, activism, or views over the scrutinized practices. As part of the recruitment strategy, individuals that were residents and/or employed in Ashburton and Glenorchy, and that were 18 years of age or older, were eligible to participate in the study. Any individual that failed to fulfill these requirements was excluded from the study. In addition, when interviews highlighted the need to engage with specific members of the community, a snowball sampling approach was employed, where the interviewer asked locals for referrals to other members of the community.

The chosen recruitment strategy enabled the capturing of individual views related to the economic, environmental, and social compatibility of water bottling and water chlorination, as well as the locals' perceptions of water-related decision-making in each location. These accounts also provided a snapshot of how and why locals decide to engage (or not) in hydropolitical conflicts over them.

Most of the data collected in Ashburton and Glenorchy was gathered through face-to-face semi-structured interviews. This format was chosen because it pays crucial attention to the respondents' narrative 'as it unfolds' (Galletta & Cross, 2013, p. 76), which in turn permits respondents to craft a story that reflects important anecdotes that may prove beneficial for the research. A semi-structured interview format also enhances flexibility in terms of structure and the ability to present further questions that may not have been included in the questionnaire guide (Mills, 2012). A semi-structured questionnaire also allows the researcher to approach respondents in casual settings and to build a rapport with them, while maintaining a basic set of themes and questions. Under such casual settings, a structured questionnaire may prove to be too rigid to allow for adequate answers from the respondents, while a fully unstructured questionnaire bears a higher risk of veering away from the themes explored in the study. A semi-structured format results in the ability to combine research rigor with spontaneity and better rapport-building between the researcher and the respondents was therefore judged to be the most suitable data collection instrument for this study.

Given the recruitment strategy explained above, this study required the ability to establish an immediate, trusting relationship with the respondents-to-be. In this regard, employing a semi-structured interview format was beneficial because it enabled a fluid, unhindered enquiry with the locals that resembled more of a conversation than an interview. To ensure the rigor of the study, a copy of the question guide was readily accessible as a point of reference when needed. Interviews were then recorded with a digital audio recorder.

In addition to the individual, face-to-face interviews, this research captured data through two focus group meetings, both of which took place in the town of Glenorchy. During the data collection process in Glenorchy, two interviewees, on two separate occasions, offered to facilitate meetings with friends, neighbors, and colleagues. These offers were agreed by the author and culminated in the organizing of two separate focus group meetings. During these meetings, the author employed the same 30-question guide that was used for the individual face-to-face interviews. During the meetings, the author made sure that everyone answered the questions, thus ensuring that the study could count each response individually.

Recognizing the controversial character of water conflicts in New Zealand and to ensure the anonymity and safety of the participants, this study took deliberate steps to minimize the traceability of their identities. These steps included replacing their names (when quoted) with a code comprised of the following formula: *Participant + unique number.* Additionally, when specific reference to their views is made, generic pronouns were used to further minimize the risk of participants being identified.

Limitations

This study's focus is devoted to examinations of value systems, traditions, personal histories, and other ambiguous issues, which as explained above, are best approached using qualitative research. This however does not come without its share of limitations, the first of which is the representative character of the study. The data collection process succeeded in capturing a total of 56 accounts: 31 from Glenorchy and 25 from Ashburton. Given the most recent census data for both towns, one cannot argue that the findings included in this study are in any way representative of the whole population in each location. Therefore, the claims emanating from this research need to be adequately confined to, and seen as reflective of, the voices of the residents only. This however does not represent a hindrance for this study, because the two proposed arguments are focused on residents' hydropolitical conflict intentionality and on the risk of intensification. The observations in this study are not meant to provide commentary of Ashburton and Glenorchy's proneness to engage in hydropolitical conflict as political units.

Secondly, this study investigates the question of hydropolitical conflict in two relatively small, rural environments. Even though the accounts and dynamics examined in this study may indeed be echoed by individuals in bigger, more urbanized contexts, this study does not provide a definitive foundation for such an assertion. This examination however does open the plausibility for researchers to explore the above dynamics in such settings. Such interrogations may shed further light on how and why hydropolitical conflicts emerge and intensify in non-rural environments.

Lastly, this study has limitations regarding the extent to which it represents Māori values and interests. This is both due to the low concentration of Māori in both locations and on unsuccessful efforts to recruit Māori participants (this is further explained in Chaps. 4 and 5). Bearing these limitations in mind, it is of crucial importance that future research examines the conditions that influence the emergence and intensification of hydropolitical conflicts within communities with higher concentrations of Māori.

Final Remarks on Research Design

In closing, this chapter presented the research agenda for this project. It did so by proposing the following research question: *How and why does the commercialization of water affect the emergence of hydropolitical conflicts?* To answer this question, this study proposed two central arguments: The first is that the introduction of water commercialization practices influences the emergence of hydropolitical conflict intentionality if the practice is economically, socially, and environmentally incompatible with local communities; second, that hydropolitical conflict risk intensifies in accordance with the extent to which residents trust the approval and appeals processes behind water commercialization practices.

To test the above arguments, this study employs a qualitative methodological approach and includes three methods of analysis: a conflict intentionality assessment, a comparative case study analysis, and a conflict party analysis. These analyses will be presented as follows. Chapters 4 and 5 will develop individual conflict intentionality analyses as they apply in Glenorchy and Ashburton, respectively. Chapter 6 will conduct a comparative case study analysis of the causal impacts of water bottling and water chlorination in each location. Lastly, the conflict party analysis will be elaborated in Chap. 7.

References

Abbas, S., Hashmi, I., Rehman, M. S. U., Qazi, I. A., Awan, M. A., & Nasir, H. (2015). Monitoring of chlorination disinfection by-products and their associated health risks in drinking water of Pakistan. *Journal of Water and Health, 13*(1), 270–284.

Apex Environmental. (2017). *Statement of intent*. Christchurch. Retrieved May 7, 2017, from www.apexenvironmental.co.nz

Ashburton District Council. (2018). *Keeping our drinking water safe*. Retrieved August 31, 2018, from https://www.ashburtondc.govt.nz/our-services/water-services/Pages/Keeping-Our-Drinking-Water-Safe.aspx

Aspers, P., & Corte, U. (2019). What is qualitative in qualitative research. *Qualitative Sociology, 42*, 139–160.

Department of Internal Affairs. (2017). *Report of the havelock north drinking water inquiry: Stage 2*. Auckland. Retrieved May 19, 2017, from www.dia.govt.nz/Government-Inquiry-Into-Havelock-North-Drinking-Water

Eidem, N. T., Fesler, K. J., & Wolf, A. T. (2012). Intranational cooperation and conflict over freshwater: Examples from the Western United States. *Journal of Contemporary Water Research & Education, 147*, 63–71.

Environment Canterbury. (2017). *Local government statement*. Retrieved June 1, 2017, from http://ecan.govt.nz/get-involved/canterburywater/

Galletta, A., & Cross, W. E. (2013). *Conducting the interview—the role of reciprocity and reflexivity. Mastering the semi-structured interview and beyond: From research design to analysis and publication*. NYU Press.

Griewald, Y., & Rauschmayer, F. (2014). Exploring an environmental conflict from a capability perspective. *Ecological Economics, 100*, 30–39.

Harris, D. (2017). *Meeting report*. Amberley. Retrieved April 19, 2017, from http://www.hurunui.govt.nz/assets/Documents/Chlorination-Files-2017/Report-to-Infrastructure-Committee-re-Water-Supply-Reticulation-2017.pdf

King, G., Keohane, R. O., & Verba, S. (1994). *Designing social inquiry: Scientific inference in qualitative research*. Princeton University Press.

Legay, C., Rodriguez, M. J., Sadiq, R., Serodes, J. B., Levallois, P., & Proulx, F. (2011). Spatial variations of human health risk associated with exposure to chlorination by-products occurring in drinking water. *Journal of Environmental Management, 92*, 892–901.

Mills, M. C. (2012). Comparative research. In L. M. Given (Ed.), *The SAGE encyclopedia of qualitative research methods* (pp. 101–103). SAGE Publications Inc.

Ministry of Justice. (2016). Meaning of the Treaty. Retrieved October 26, 2017, from https://www.waitangitribunal.govt.nz/treaty-of-waitangi/meaning-of-the-treaty/

Muru-Lanning, M. (2012). Māori research collaborations, Mātauranga Māori science and the appropriation of water in New Zealand. *Anthropological Forum, 22*(2), 151–164.

Newshub. (2016). *Hundreds protest against Ashburton water sale*. Retrieved February 23, 2018, from http://www.newshub.co.nz/home/new-zealand/2016/04/hundreds-protest-against-ashburton-water-sale.html

Poata-Smith, E. S. T. A. (2004). The changing contours of Māori identity and the Treaty settlement process. In J. Hayward & N. Wheen (Eds.), *The waitangi tribunal* (pp. 165–183). Bridget Williams Books.

Radio New Zealand. (2017). *Millions of liters of artesian Otago water for sale*. Retrieved July 26, 2018, from https://www.radionz.co.nz/news/national/326929/millions-of-litres-of-artesian-otago-water-for-sale

Ragusa, A. T., & Crampton, A. (2016). To buy or not to buy? Perceptions of bottled drinking water in Australia and New Zealand. *Human Ecology, 44*(5), 565–576.

Rifkin, E., & Bouwer, E. (2007). *The illusion of certainty: Health benefits and risks*. Springer.

Statistics New Zealand. (2018a). 2013 *Census—QuickStats about a place: Glenorchy*. Retrieved March 8, 2018a, from http://archive.stats.govt.nz/Census/2013-census/profile-and-summary-reports/quickstats-about-a-place.aspx?request_value=15005&tabname=Business

Statistics New Zealand. (2018b). 2013 *Census QuickStats about a place: Ashburton District*. Retrieved March 7, 2018b, from http://archive.stats.govt.nz/Census/2013-census/profile-and-summary-reports/quickstats-about-a-place.aspx?request_value=14909&tabname=Business

Steinberg, P., & VanDeveer, S. (2012). Comparative environmental politics in a global world. In P. Steinberg & S. VanDeveer (Eds.), *Comparative environmental politics: Theory, practice, and prospects*. MIT Press

Stuff. (2013). *Ngai Tahu follows the water*. Retrieved May 14, 2019, from http://www.stuff.co.nz/business/industries/9200166/Ngai-Tahu-follows-the-water

Stuff. (2016). *Elation as Ashburton council backs out of controversial water bottling deal*. Retrieved May 22, 2019, from https://www.stuff.co.nz/business/81873688/ashburton-council-backs-out-of-controversial-water-deal

Water New Zealand. (2018). *Water New Zealand welcomes Christchurch water chlorination decision*. Retrieved April 19, 2019, from https://www.waternz.org.nz/Story?Action=View&Story_id=702

Chapter 4
Glenorchy

Abstract This chapter examines the proposed arguments in the town of Glenorchy, located in the western part of New Zealand's Otago region. It finds evidence suggesting that incompatibilities over the introduction of water chlorination were more salient than over water bottling. The escalation of hydropolitical conflict risk is further explained through skepticism over the decision-making process behind water chlorination.

About Glenorchy

The town of Glenorchy is located within the jurisdictional authority of the Queenstown-Lakes District Council, in the western part of the Otago region. Figure 4.1 shows an aerial view of the town (circled in black).

There is an isolation of sorts that dominates life in Glenorchy; as follows from the picture above, it is easy to note that the town is enclosed by different geographical elements. The town is bordered by the massive Lake Wakatipu in the south, by the Rees and Dart river systems in the west, and by the Southern Alps in the north and east, effectively limiting its direct access to the rest of the country to one road in the southeast. During the data collection, this isolation manifested in several ways, as will be explained later in this chapter.

While engaging with locals, their predilection for the Status Quo became clear, which confirmed some insights I had already identified elsewhere. For example, in 2012 a project was proposed that would have developed an 11-km tunnel connecting Glenorchy to Milford Sound, in the bordering region of Fiordland (Stuff, 2012). The tunnel's intent was to shorten the driving distance between Milford Sound and Queenstown, two important tourism centers in the South Island. It was expected that the tunnel would have dramatically increased traffic coming through Glenorchy, as well as the influx of tourists requiring services in the town. After receiving massive opposition from the locals, the project was abandoned.

During the identification of potential case studies, I was informed of the well-known reputation of Glenorchy residents, and how they seemed to espouse a type of locally grown pride in their town. This pride was evident in their use of the term

A. E. Suazo, *The Risk of Water Conflicts in Aotearoa-New Zealand*,
https://doi.org/10.1007/978-3-031-06660-3_4

Fig. 4.1 An aerial view of Glenorchy

'Head of the Lake' when referring to Glenorchy, the installation of a welcome sign that reads: 'Welcome to Glenorchy—Gateway to Paradise', and their celebration of Glenorchy's contributions to the war effort through scheelite mining and provision of soldiers.

According to Statistics New Zealand, Glenorchy has a population of 363 people, comprised by an almost equal amount of female and male respondents: 183 and 180, respectively (2018). Additionally, slightly over 70% of Glenorchy's population was found to fall between the ages of 15 and 64 years, with a median age of 39.2 years (Statistics New Zealand, 2018). Working residents were employed in several different occupation categories including managerial positions (26%), professional trades (14%), technicians and trade workers (15%), community and personal service workers (15%), clerical and administrative workers (5%), sales workers (5%), machinery operators and drivers (7%), and laborers (13%) (Statistics New Zealand, 2018). Additionally, an estimated 86.6% of Glenorchy's citizens aged 15 years and over had completed some type of formal qualification (Statistics New Zealand, 2018).

In terms of the town's ethnic makeup, it is difficult to ascertain the real ethnic belonging of Glenorchy's residents, because Statistics New Zealand's coding system allows for citizens to input more than one ethnic background, which allows the total percentage of ethnic calculations to surpass the 100% mark (Statistics New Zealand, 2018). Notwithstanding the above coding issue, Statistics New Zealand reports that most of Glenorchy's inhabitants are of European descent (96.2%), 11.3% were Māori, 1.9% reported being Pasifika, 1.9% were Middle Eastern, Latin American or African, and 1.9% were of other ethnic backgrounds (2018).

About the Participants

It is within the above-described context that the data collection process was established: It took place between January 17 and 24, 2018, at which point 31 respondents

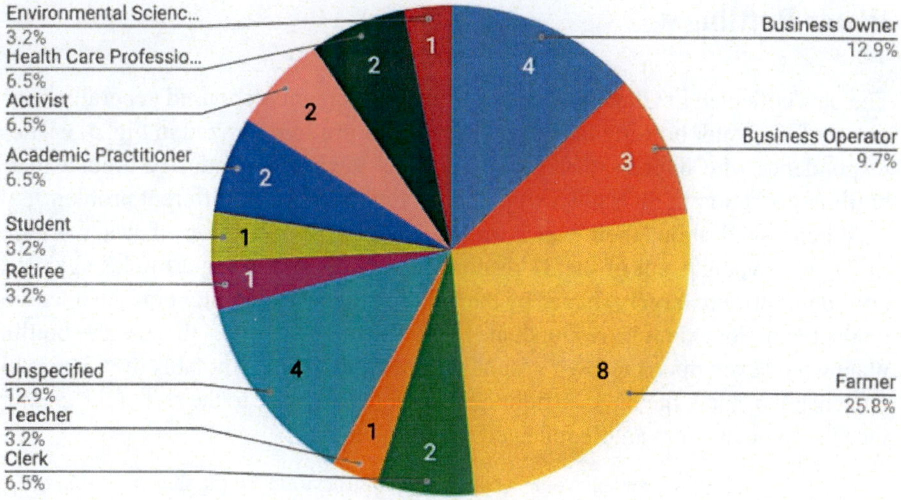

Fig. 4.2 Participants per occupation in Glenorchy

took part in the study. Participants reported being involved in a wide range of occupations, which are summarized in Fig. 4.2. Even though the greatest respondent cohort reported coming from the farming sector, the distribution of participants relative to occupation is fairly even. This provides an ample reflection of the potential incompatibilities that water bottling and water chlorination could exert upon a broad range of interest groups in Glenorchy.

In addition to occupation, this study successfully captured the voices of a wide range of age groups and ensured a comparable representation along gender lines. Out of the 31 participants that took part in this study, 19 were female and 12 were male. With regard to age distribution, 35.5% of the respondents reported being 60 years old or beyond, while 32.3% reported ages between 18 and 40. Additionally 12.9% of the respondents fell between the ages of 40–60, while 19.4% did not wish to share their age. Considering the town's age distribution, this study captured sufficient voices in each age bracket to reflect the participants' views and attitudes over water along age lines.

In terms of ethnicity and immigration status, the great majority of participants from Glenorchy, 23, reported being of European descent, and a small fraction, 8, reported being first generation immigrants. New immigrants reported coming from Hong Kong, the UK, the USA, Australia, and Germany. This distribution corresponds to the latest census information available for Glenorchy, where most of the town's residents are of European background, while a combined 5.7% are regarded as Pasifika, Asian, Middle Eastern, Latin American, African, and other (Statistics New Zealand, 2018).

It is noted that no participant of Māori descent was identified during the data collection process, making this analysis unrepresentative of Indigenous views and positions vis-a-vis freshwater and their engagement in hydropolitical conflict in Glenorchy.

Water Bottling

The data collected in Glenorchy suggests that participants espoused generally mixed views of the water bottling industry. These views are summarized in Fig. 4.3. Those respondents who expressed negative attitudes toward the industry argued that a bottling operation in their town would generate a number of different problems.

When asked about their views on the potential introduction of water bottling in Glenorchy, eight out of the 31 respondents expressed strong attitudes against it, attitudes that were largely shaped by the indirect impacts of a water bottling operation on the town. For example, respondents highlighted the pressures that a water bottling plant would exert upon the town's roading infrastructure, most notably the Glenorchy-Queenstown Road, the only corridor connecting Glenorchy to the rest of the country. One respondent for example reflected the following:

> Probably the biggest problem would be the amount of trucks around the area. It wouldn't be the water per se. If it is a commercially-viable thing, then you would be seeing 10 tankers a day. To be commercially-viable, they're not going to take 100 L of water. They'll take 100 thousand liters of water. (Respondent 6)

Other respondents expressed worries over increased noise pollution and traffic, and over farmers' continued ability to transport their produce to market. Additionally, concerns over the plant's potential impact on the proliferation of plastic waste further bolstered some participants' dislike for the prospect of a water bottling operation in their town. For example, one participant explained their dislike toward water bottling as follows:

> If everyone in the community starts drinking bottled water, how are you going to get rid of the plastic in the bottles? It has all sorts of consequences, which is not about protecting the health of the community. (Respondent 11)

However, the water bottling operation did not seem to be widely rejected in Glenorchy presumably because of the abundance of water in the periphery. In this regard, the data shows that nine participants expressed some type of acceptance for

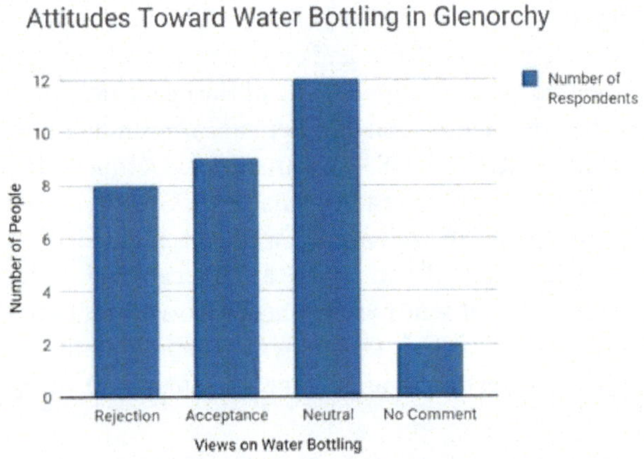

Fig. 4.3 Attitudes toward water bottling in Glenorchy

the installation of a water bottling operation in their town, a respondent cohort that was slightly higher than that which expressed a rejection for the operation. Water bottling did not seem to antagonize the town's respondents, or business interests. On this note, one respondent, who operates a local hotel, argued:

> Water bottling is an activity that'll happen, and if it is done well, it will happen, and hardly anyone will notice what is going on. (Respondent 7)

When explaining whether water bottling in Glenorchy was problematic, a second respondent, employed in the tourism industry, reported:

> It depends on the impact [of the project]. You can extract as much water from this lake [Wakatipu] as you want and have no impact on anyone. Whether that business would contribute one way or another, while taking the resource, is a different point. (Respondent 2)

A third respondent, who operates a local environmentally friendly campground, was unaware of the water bottling operation and seemed more concerned with developing a business that fostered awareness about water consumption and use, than with their business or interests being challenged by a highly extractive operation in the vicinity.

Based on the above accounts the water bottling operation does not seem to be openly rejected by the respondents, but support was contingent on a series of conditions; the most prominent is a commitment to environmental responsibility by the company in question. For example, one respondent argued:

> If there is some kind of negative impact, that should be paid for by the commercial operation. And that's got to change depending on what the source of the water is. I have a fairly firm view that if someone takes water right before it goes to the sea, what's the issue? […] I guess what I am saying is, if there is a potential downstream impact, then [the project] needs a lot more consideration than if there isn't. (Respondent 6)

The respondents' primary concern regarding the water bottling operation was related to its potential impact on water quality degradation in their town. In this regard, one respondent expressed their concerns, saying:

> For me, the thought of people bottling water up there, my concern is potential contamination downstream, and also potential impact on my ability to take water from my bore, as it is downstream from there. (Respondent 2)

Some respondents also expressed that they would support the water bottling operation if it benefited the town. However, these respondents were not clear about what specific benefits would satisfy them, although initiatives such as infrastructure investment, funding for community events, job creation, and cultural promotion of the town were points of convergence. Respondents from Glenorchy promote the accomplishments of their locals, which helps explain why approval for anyone commercial operation is best facilitated through support for the cultural and historical promotion of the town. In fact, some respondents documented cases where commercial transactions obstructed the celebration of the town's history, which resulted in some type of discontent in the community. One respondent, for example, said the following:

> There is an old stamp battery up the hill, that was traditionally for scheelite mining, about 100 years or more, and needed water. So, there was a storage pond that was all water driven. The property surrounding it was bought by an American and stopped all access. That kind of thing annoys me. Because he could've said: It is not commercially used now but carry on doing it. It is not commercially used, but it can be a demonstration of what old fellows used to do, but he stopped the access. (Respondent 6)

Alongside accounts that expressed either acceptance or rejection for water bottling, the data also shows a striking number of participants—14 in total—reporting having no position whatsoever vis-a-vis water bottling in Glenorchy, or not wishing to make any comment about it. The motivation behind these views was largely shaped by two factors, the first of which was respondents not knowing about the water bottling project, and second, that they did not seem to consider water bottling as affecting them negatively in any way.

In addition to mixed views over water bottling as a significant cause of water concerns in Glenorchy, respondents did not express major distrust toward the approval process behind the operation. In fact, some participants expressed an unwillingness to interfere in what they regarded was a private transaction between the previous owner of the property where the bottling plant is expected to operate, and the bottling company. There are no indications in the data that suggests that the approval and appeals process behind Koha Water Limited's proposed operation was dubious or untrustworthy. Records from the ORC, the water authority responsible for the conferring of consents for bottling, document the degree of oversight it exerted in informing the public about the bottling project. In its assessment, the ORC took a keen interest in detailing the precise nature of the project, including a full study of the operation's point of extraction, its impact on river flow levels, and an effort to explain the extraction rate of the bottling operation relative to other activities. On this last note, the ORC reported the following:

> The Koha Water take is a small water take and is equivalent to the water used by a small irrigated farm. Typical farms in Otago using irrigation would use 6,000,000 L per hectare (Aqualinc) per season plus water for stock use. On this basis the amount of water taken by Koha would be equivalent to a farm of less than 40 hectares. (Otago Regional Council, 2017, p. 17)

Bearing the above observations in mind, it is difficult to regard the approval and appeals process behind Koha Water Limited's proposed bottling operation as doubtful, or as inspiring skepticism by the respondents. That, however, does not mean that they unanimously agreed with the operation; it means that those who disagreed with the introduction of water bottling were likely to seek access to the ORC's appeals process to address any given grievance or incompatibility.

Water Chlorination

The data suggests that respondents in Glenorchy strongly disliked water chlorination, a policy seen by most participants as a significant source of water worry or insecurity in the town. From the 31 respondents who took part in this research, 28 expressed significant opposition toward the chlorination scheme, many of whom evoked deep concerns about four central themes: environmentalism, public health, local suitability of the policy, and water-related decision-making practices. These expressions of dislike were also shared across occupations, age groups and gender. In addition to this, one respondent expressed neutral views over chlorination, and two more did not make any comments over the policy. In Glenorchy, no respondent expressed support or approval for water chlorination (Fig. 4.4).

In terms of the impact of water chlorination on Glenorchy's ecosystem, respondents expressed significant concern over how the chlorine in the water could potentially affect the volume of flora and fauna that exist in the immediate vicinity of the town. For example, one respondent highlighted their worries over the impact of chlorine on Lake Wakapitu's fish stock, which relates to the lake's key role in the sustenance of populations of eel, trout, salmon, and several species of Indigenous fish threatened with extinction (Otago Regional Council, 2016a), many of which sustain several types of recreational and subsistence fishing activities in Glenorchy. Another respondent conveyed their worries over chlorination by stressing that chlorinated water could impose a negative impact on the town's ability to nourish its crops and, therefore, on its ability to adequately address its own food security needs.

A respondent who works as a corporate consultant argued that part of the problem sparked by chlorination emanates from a deep lack of systemic thinking by the QLDC. In this regard, they said the following:

> In my experience, it is not just residents: It is Council. It is people that are not taught to think systematically. So, I think it is bigger than the individuals, in any of the groups you are working with. We are incentivized to give answers in school, we are not asked to ask questions. And as we get older and things become greyer, it is unsurprising that we take that approach, with farmers having one view, residents having another one [...] because they have those motivations, and it is uncommon for someone to ask us to step out of that box. (Respondent 30)

Along with issues related to environmental concerns, respondents expressed worries over the causal connection between chlorine exposure and cancer. For

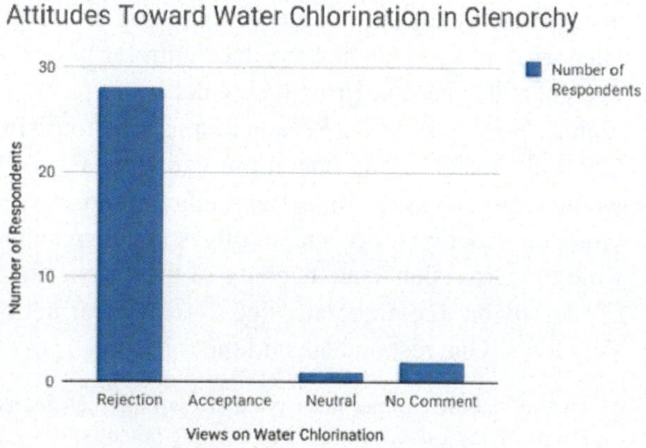

Fig. 4.4 Attitudes toward water chlorination in Glenorchy

instance, a respondent employed in the public health sector registered unease over this connection, a worry that was echoed by two other respondents, themselves cancer survivors who relocated to Glenorchy to have access to better environmental conditions. Other participants reported having had stomach problems and discomfort shortly after drinking chlorinated water. One participant said the following:

> Before [chlorination], drinking water from the tap was no problem at all. I didn't feel any problem at all. But after they put chlorine, I drink water from Glenorchy and I feel sick. (Respondent 8)

Other respondents expressed their dislike for water chlorination by emphasizing the water's taste. On this issue, one respondent said:

> I haven't drank out of the tap since [chlorination] started and have been boiling water until we can afford to get a filter, because the water out of the tap tastes foamy and thick, and a little bit soapy. You can taste it. (Respondent 5)

Evoking similar health-related concerns, another respondent argued that water chlorination was an issue that affected several dimensions of residents' health. They said:

> It is not just about drinking water. The skin is the biggest organ we have, and I get in the shower every morning, and it is chlorinated water that is getting through my skin, so we are ingesting it in different ways. I think drinking water gets too much attention, and it goes back to that systemic view: [Chlorine] is coming back to us in other ways that we don't immediately recognize. (Respondent 30)

Departing from health concerns, another respondent related to the chlorination policy to other activities in the district, which together may be contributing to the deterioration of the town's water supplies. They said:

> We do not understand aquifers: The size of them, how much water they have, what systems they recharge, their recharge rate etc. When you focus on one thing such as chlorine, you can argue for or against, but you need to look at the whole system. There are V8 boats going back and forth the river, causing a series of problems. All these issues are part of the same problem. You need to look at the whole and get a big picture. (Respondent 12)

According to the data, the water chlorination decision was met with further disapproval by the community, in part, due to its dubious suitability for Glenorchy. Glenorchy is a jurisdiction that has, until fairly recently, eluded the intensive development and economic growth to which other parts of the country have grown accustomed. Notwithstanding seasonal spikes in tourism activity, the town has mostly low-scale businesses in operation, with very few of the amenities and services one would expect to see in bigger and more urbanized settings. When asked about their views on water quantity and quality, some respondents argued that both water traits were in an excellent state because of the absence of highly abstractive industries in Glenorchy or, to extrapolate, due to freshwater not being overly commercialized in their town. One respondent said the following:

> History has shown that intensive dairy farming has degraded water quality. So yes, it has the potential. I don't think it has a strong potential the way they farm the land here. Their

> stocking rates are much lower than other parts of the country because the production is low, and the land is not as good. We don't have that much flat land [...] like the Canterbury Plains [...] so the risk is lower here because of that. (Respondent 2)

Another respondent also alluded to the absence of dairy farming when describing the high quality of Glenorchy's water but argued that this perception may be changing in view of the growth of other industries such as tourism. They said:

> I think here we are not bad, because we don't have any dairy. And we don't, it hasn't affected us yet. But like in Glenorchy they've had these scares over e-coli this year with their water, because their town is growing so much, and it has changed the way in which they treat their water. It has never been updated to keep up with the growth, and suddenly they have all of these issues, because there are more people to deal with. (Respondent 10)

Another respondent argued that despite some presence of e-coli particles in the water, the risk of industry and farming-related water contamination was minimal, which rendered chlorination of the town's water supply unnecessary in their view. This respondent's elaboration finds some support in a 2006 report by the ORC, which assessed the water quality of the towns of Kingston and Glenorchy. The report documents that between 1996 and 2003, Glenorchy's public water supplies showed no signs of fecal coliform bacteria, and minimal traces of nitrite-nitrate nitrogen (NNN) (Otago Regional Council, 2006) and points out that direct land use posed no immediate risk upon the town's water take (Otago Regional Council, 2006). Further testing during the period between June 2011 and June 2016 shows low traces of pathogens and NNN (Otago Regional Council, 2016b), proving both the benign state of the town's freshwater, and the questionable need for chlorination in some participants' views.

Lastly, the data gathered in Glenorchy suggests that dislike for water chlorination was also driven by disagreements over how the decision to chlorinate came about. For some respondents, the chlorination policy was motivated by an attempt by the QLDC to avoid liability in cases of waterborne diseases. One respondent, for example, said the following:

> It is very real for all of us, and one of the things that irritates me is that Council is like someone taking antibiotics: they can go damage themselves without getting an infection, instead of not allowing the damage to happen, and stop taking drugs. That's what they have done. Chlorination allows them to be as slack as they want. (Respondent 18)

Based on several respondents' comments, the QLDC conceived a water securitization strategy that did not adequately account for local values, and that overlooked meaningful deliberation with the town's local experts. From the two focus group meetings held for this study, comprising a total of 16 respondents, there was a consorted view that proper consultation was not undertaken by the QLDC with regards to the implementation of the recommendations of the Havelock North Report. One respondent, for example, said the following:

> They [QLDC] haven't looked at options. They already decided what they wanted, and made the process fit. We are waiting to make a decision on what action to take, because now apparently there is some evidence that suggests we should chlorinate because the reticulations

> have been so under-cared. But we've gone to the ombudsman, and other organizations. An auditor came in to assess the [water safety] plan, and said it was all ok, so we agreed to it, and that was seen as enough consultation with the community. (Respondent 19)

One participant argued that the lack of consultation may have been caused by the QLDC favoring outside expertise over the local one. In this regard, the respondent said the following:

> Council doesn't discuss [chlorination], because they don't understand it, and they don't want to discuss it. They want to trust their managers. They say we're not experts, [that] we don't know, but these people are. (Respondent 16)

Some respondents argued that, had proper consultation been started in Glenorchy, simpler and more cost-effective solutions would have been recommended by the locals, which would have led to less intrusive prevention strategies. One respondent, for example, noted that fears over water contamination could have been mitigated easily by building a fence around the town's water bore, thus preventing farm animals from approaching and negatively affecting the town's groundwater source, a practice that aligns with the principles behind the multiple barrier approach for drinking water safety (Ministry for the Environment, 2018). A decision-making process that was perceived as unilateral by the respondents exhausted an opportunity to launch a risk mitigation strategy both palatable to residents, and which could have catered to their specific requirements without breaching or imperiling their own water-related values and interests.

Some participants also reported having concerns with the precedent that chlorination could create in their district. This concern was related to the involvement of the private sector in the delivery of water services. One respondent said the following:

> [Commercialization] is part of that picture, because for the sake of reducing liability, it is easier to involve a private enterprise. There is no doubt in my mind that Council will end up selling their water infrastructure, and have private companies control it, like they did in Auckland. (Respondent 19)

Another participant expressed worries over such a scenario, because in their view, the private provision of water services would imperil the community's consultation rights.

Results

When accounting for the type of water-related practice that respondents regarded as being the most incompatible to their interests, participants seemed more likely to exhibit conflict intentionalities over water chlorination than over water bottling: from the 31 respondents who participated in this study, 23 reported an intent to engage in some type of hydropolitical conflict action against water chlorination. Out of these accounts, 11 documented being inclined to engage in intense hydropolitical conflict through litigation. Additional actions were documented at different parts of the CCI

Scale, with two respondents willing to engage in petitioning, and 10 expressing their will to voice discontent through their elected officials. In addition to this, eight participants expressed unwillingness to take any action. The data therefore indicates that more respondents reported intense hydropolitical conflict intentionality over chlorination than otherwise, although a comparable amount of voices stressed their will to participate in what is considered to be low-intensity conflict action (Fig. 4.5).

The data suggests that the number of respondents willing to engage in conflict events of high intensity was comparable across gender lines. For example, from the 11 respondents who reported an intent to engage in litigation procedures, five were female and six were male. The data, however, shows that there is a clear imbalance with regard to the intent to engage in low-intensity conflict events over water bottling and water chlorination relative to gender. For example, nine female respondents reported being willing to voice their discontent over water bottling and/or water chlorination through their elected officials, compared to one male participant. In addition to this, the two respondents that reported being willing to become involved in petitioning were female, with no male cohort expressing a similar intent to engage in such activities. In this regard, the data shows little variation of high-intensity conflict engagement between female and male respondents, but the former cohort seemed more likely to exhibit low-intensity conflict intentionality in response to water bottling and/or water chlorination. The gender distribution related to conflict intentionality is presented in Table 4.1.

Some of the documented expressions of conflict intentionality against water chlorination in Glenorchy have translated into a number of manifested conflict events. For example, official QLDC meeting minutes document that respondents from Glenorchy attended at least three public meetings between December 2017 and March 2018

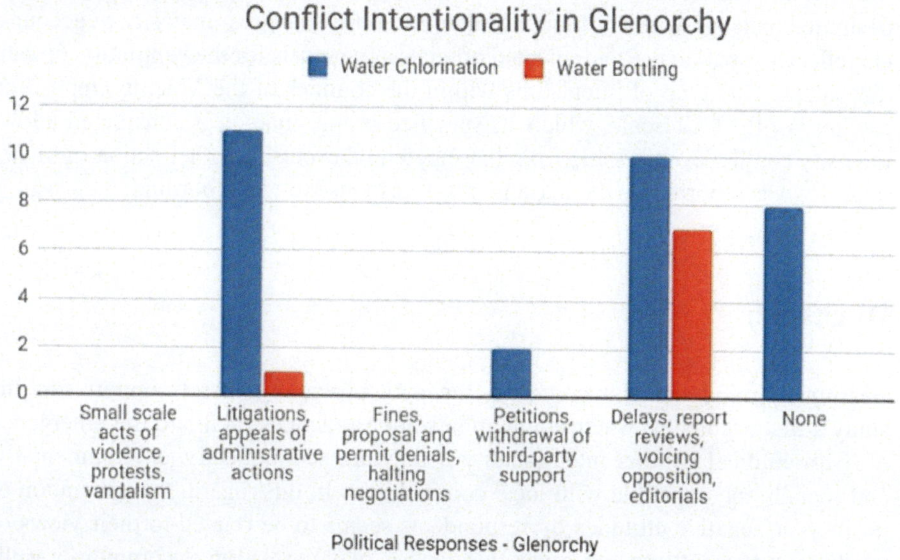

Fig. 4.5 Distribution of hydropolitical conflict intentionality in Glenorchy

Table 4.1 Gender distribution of hydropolitical conflict intentionality in Glenorchy

Type of Action	Female respondents	Male respondents
Small scale acts of violence, protests, vandalism	0	0
Litigations, appeals of administrative actions	5	6
Fines, proposal and permit denials, halting negotiations	0	0
Petitions, withdrawal of third-party support	2	0
Delays, report reviews, voicing opposition, editorials	9	1
None	3	5

to voice their discontent toward water chlorination (Queenstown-Lakes District Council, 2017, 2018a, 2018b). During one of these events, one community member poured water on the Council floor as a sign of protest, which culminated in the individual being removed from the Council premises (New Zealand Herald, 2017). Since then, further lobbying and organizing around the chlorination issue have occurred in Glenorchy, with continued threats to intensify the ongoing conflict. During the data collection, a local organization by the name of Sustainable Glenorchy expressed that they had already sought legal counsel to begin a legal challenge in court over chlorination. In addition to this, interviews with some of the respondents expose their plans of marching along the Glenorchy-Queenstown Road, which would arguably cause significant disruption of economic activities dependent on that road, namely tourism and the transportation of produce to market. Such protests have not come to pass, but nevertheless suggest there is a palpable risk of hydropolitical conflict intensification.

Unlike water chlorination, water bottling inspired mostly low-intensity hydropolitical conflict intentionality. With the exception of one respondent who expressed a desire to challenge water bottling through the courts, the great majority of accounts (7) reflected a will to use their elected officials as channels for the ventilation of their discontent. This type of intent falls within the confines of the Voicing Opposition category of the CCI Scale, which, as specified in this chapter, is considered a low-intensity conflict event. Aside from these potential events, no other action of higher intensity was reported by the participants in response to water bottling.

Discussion

Bearing in mind the above results, the data provides general support for this study's first argument: water commercialization practices facilitate the emergence of hydropolitical conflict intentionality if they are economically, environmentally, and socially incompatible with local communities. In this regard, the generation of positive or negative attitudes by respondents seems to be related to their views of what part water bottling and water chlorination play in relation to community well-being. In most cases, those individuals that reported having strong views against water chlorination argued that the policy imperiled their own ability to benefit from freshwater access in some way, irrespective of occupation. In contrast, respondents seemed less likely to regard the prospective water bottling operation as incompatible in their town, in part due to the abundance of freshwater found in Glenorchy, and to the potential benefits of such an enterprise.

The great majority of participants in Glenorchy espoused negative attitudes toward water chlorination, while holding mostly neutral views on water bottling. It must be noted, however, that the second-largest concentration of participants seemed to approve of the water bottling industry, albeit with a series of conditions attached to their approval. Unlike chlorination, water bottling did not inspire strong negative attitudes in Glenorchy.

The data supports the second argument proposed in this study, namely that hydropolitical conflict risk intensifies when communities do not trust the approval and appeals processes behind a proposed water commercialization activity. In Glenorchy, this argument was evident in the community's lukewarm intent to engage in intense hydropolitical conflict over water bottling. This does not entail that conflict over water bottling was absent; it means that conflict intentionality over water bottling was distributed in the lowest levels of the CCI Scale, where hydropolitical conflicts are ventilated through the institutional and procedural purviews of local councils. When trust in these processes was compromised, as with the introduction of water chlorination in Glenorchy, respondents tended to espouse significant disdain toward the water authority system and were likely to report a preference for alternative avenues and processes to sway decision-making in their favor.

The data indicates that Glenorchy's respondents seemed to be confronted with two crucial moments vis-a-vis their participation in hydropolitical conflict. The first relates to their views and perceptions of the generation of incompatibilities concurrent with water bottling and chlorination. Those respondents who regarded either one of these practices as deleterious were then confronted with a second crucial moment related to the intensity of their potential political action. In this regard, concerns over chlorination or bottling motivated the emergence of hydropolitical conflict intentionality at various levels, but the data indicates that the risk of conflict intensification seemed higher when respondents did not trust the approval and appeals processes behind each operation. Therefore, the risk profile for intensification seemed more likely in the face of water chlorination than water bottling.

While benign in spirit, the question of water securitization through treatment has not been universally embraced by all districts in New Zealand, and Glenorchy's ongoing concern over chlorination demonstrates the extent to which a community is likely to resist such policies, and the rationale that drives their resistance. To some degree, the QLDC's decision to chlorinate the town's water reflects an acute inattention to (or disregard for) the non-commercial and systemic bond that some respondents in Glenorchy have established with their local environment. Based on the accounts by those respondents who have been politically active against chlorination, ideas such as local ecosystemic awareness and the importance of protecting sources of drinking water seem to clash with the QLDC's perception that water security can only be attained via treatment. It can be surmised that the QLDC assumed, perhaps wrongly, that chlorination would be met with support from the community due to the policy's ability to securitize the town's water from any potential risk. This policy, however, seemed ill-conceived for a town like Glenorchy, where no major justification for treatment had been documented.

Also related to the question of what drives water-based conflict (over quality or quantity), the decision by the QLDC to chlorinate Glenorchy's freshwater seems to impact the town's self-reliance (such as local food production), and its demands for health services. These points carry additional weight, considering the geographical location of the town, and the products and services sustaining it. Glenorchy lies on the northern shores of Lake Wakatipu and is enclosed by the Southern Alps, effectively separating the town from the rest of the QLDC's jurisdiction. The nearest fully serviced center, Queenstown, lies roughly 46 kms south of Glenorchy, which partly explains some of the respondents' strong resistance to the chlorination scheme. In a town with only one convenience store, local forms of subsistence food production and procurement are of extreme importance for Glenorchy's residents, and any policy detrimental to such an enterprise is likely to be faced with resistance.

Additionally, the visible lack of public health services in Glenorchy raises questions as to the ability of residents to receive proper medical care should a chlorine-related concern arise. Prior to the installation of the chlorination scheme, medical services in Glenorchy were already limited, and remain so today: healthcare services in the town are delivered by one volunteer nurse who is available to treat patients once a week for three hours, in premises that are shared with the local library and museum (Glenorchy Community Association, 2019). Whether water chlorination will increase demands for health services remains open for debate and is an issue that falls beyond the scope of this study. It does, however, raise questions related to respondents' perceptions that their health-related needs are being influenced by a policy that already antagonizes other aspects of their lives. Those respondents who questioned the decision to chlorinate on the grounds of its suitability for the town seem likely to believe that any demand resulting from the chlorination scheme is fabricated, and thus likely to fragilize the community's ability to adequately address its own needs. This is not to say that the QLDC had the explicit intent to artificially generate the above incompatibilities: it nevertheless demonstrates how and why the water chlorination decision is seen as incompatible with the community's local fulfillment of needs.

Finally, the intensification of hydropolitical conflict risk is explained by accounting for the approval process behind the water chlorination policy. As per the elaboration above, chlorination was already regarded as incompatible with Glenorchy's water security framework, and as likely to produce water-based divisions in the town. The QLDC's perceived unilateral approval of the policy, and the limited time allotted for consultation with the town's residents, worsened the respondents' overall orientation toward the policy. Therefore, one can argue that Glenorchy's hydropolitical conflict risk profile and subsequent organization against water chlorination were driven by perceived problems resulting from the chlorination policy, and the respondents' discontent toward the approval process itself. Had proper consultation been established in Glenorchy, it would have perhaps left space for the respondents to advance alternative solutions from within the water authority system, thus reducing the risk of intensification of conflict dynamics. This under no condition negates the QLDC's drive to chlorinate Glenorchy's water, as the issue seems to have been treated as a matter of urgency, with little time to deliberate on the applicability or implementation of the policy. Glenorchy's resistance toward chlorination, however, demonstrates the challenges that local councils in New Zealand face in their attempts to securitize the freshwater under their jurisdiction, and the extent to which communities are willing to organize politically against these types of policies.

References

Glenorchy Community Association. (2019). *Nurse clinics*. Retrieved May 23, 2019, from http://www.glenorchycommunity.nz/community/community-services/nurse-health/

Ministry for the Environment. (2018). *Technical guidelines for drinking water source protection zones*. Christchurch. Retrieved January 27, 2022, from https://environment.govt.nz/assets/Publications/Files/technical-guidelines-for-delineating-drinking-water-source-protection-zones.pdf

New Zealand Herald. (2017). *Woman makes bizarre Ed Sheeran protest at Glenorchy water chlorination meeting—NZ Herald*. Retrieved October 12, 2018, from https://www.nzherald.co.nz/nz/news/article.cfm?c_id=1&objectid=11960357

Otago Regional Council. (2006). *Groundwater quality in Kingston and Glenorchy*. Retrieved August 6, 2018, from https://www.orc.govt.nz/media/3818/kingston-and-glenorchy-groundwater-quality-web.pdf

Otago Regional Council. (2016a). *Regional plan: Water for Otago*. Retrieved August 8, 2018, from http://archive.orc.govt.nz/Documents/Publications/Regional/Water/PlanChange6A/2016/RegionalPlanWater-UpdateFeb2016a.pdf

Otago Regional Council. (2016b). *Where is the flow site for the Glenorchy area?* Retrieved August 27, 2018, from https://www.orc.govt.nz/media/4303/2016-wq-report-card-glenorchy.pdf

Otago Regional Council. (2017). *Otago regional council meeting 5 April 2017*. Dunedin. Retrieved May 27, 2019, from https://www.orc.govt.nz/media/2338/combined-public-agenda-for-council-meeting-5-april-2017.pdf

Queenstown-Lakes District Council. (2017). *Queenstown Lakes district council—14 December 2017*. Queenstown. Retrieved June 1, 2019, from https://www.qldc.govt.nz/assets/Uploads/Council-Documents/Full-Council-Meetings/2018/8February2018/0.-Confirmation-of-minutes-of-14-Dec-2017.pdf

Queenstown-Lakes District Council. (2018a). *Queenstown Lakes district council—8 February 2018*. Queenstown. Retrieved June 1, 2019, from https://www.qldc.govt.nz/assets/Uploads/Council-Documents/Full-Council-Meetings/2018/8February2018a/0-Agenda-summary.pdf

Queenstown-Lakes District Council. (2018b). *Queenstown Lakes district council—23 March 2018*. Queenstown. Retrieved June 1, 2019, from https://www.qldc.govt.nz/assets/Uploads/Council-Documents/Full-Council-Meetings/2018/3-May-2018b/0-Council-mins-23-March-2018.pdf

Statistics New Zealand. (2018). *2013 Census—QuickStats about a place: Glenorchy*. Retrieved March 8, 2018, from http://archive.stats.govt.nz/Census/2013-census/profile-and-summary-reports/quickstats-about-a-place.aspx?request_value=15005&tabname=Business

Stuff. (2012). *Glenorchy says no to tunnel*. Retrieved March 5, 2022, from https://www.stuff.co.nz/southland-times/6302082/Glenorchy-says-no-to-tunnel

Chapter 5
Ashburton

Abstract This chapter examines the conflict-inducing potential of water bottling and water chlorination in the town of Ashburton, located in the Canterbury region of New Zealand's South Island. Evidence presented in this chapter suggests that the introduction of water bottling presented incompatibilities that were further exacerbated by negative views over the approval process assessing the project.

About Ashburton

The town of Ashburton is located 89 kms southwest of the city of Christchurch, in the Canterbury region of New Zealand's South Island. Ashburton is administered by the Ashburton District Council, whose jurisdictional authority is confined within a combination of geographic elements. Ashburton is also under the greater regional authority of Environment Canterbury. Figure 5.1 shows an aerial view of the town (circled in black), in relation to the rest of the district. The area of authority of the Ashburton District Council is demarcated by the flows of the Rangitata and Rakaia rivers in the western and eastern parts of the district, by the Southern Alps in the north, and by the South Pacific coast in the south.

On more than one occasion, during the data collection process respondents described their district as an island of sorts. Excluding the Southern Alps to the north, the town of Ashburton (and the district as a whole) is mostly contained (and even fragmented) by bodies of water. In discussing water issues with locals, respondents shared insights into the impacts of flooding events in the area during the winter months, which often leads to the closing of crucial roading infrastructure, supply chains, and general damage. They also expressed anxiety over the district's drought-proneness during the summer months, where water restrictions seemed commonplace. These accounts painted a picture of a town caught between two extremes, each one affecting residents differently.

Most of the town of Ashburton's population is concentrated in six subdivisions: Ashburton Central West (1035), Ashburton Central East (1722), Hampstead (2736), Allenton West (2052), Allenton East (4131), and Netherby (1821), which together hold a population of roughly 13,500 people (Statistics New Zealand, 2018). Out of

A. E. Suazo, *The Risk of Water Conflicts in Aotearoa-New Zealand*,
https://doi.org/10.1007/978-3-031-06660-3_5

Fig. 5.1 An aerial view of Ashburton

its total population, roughly 48.3% of Ashburton's residents are male and 51.7% are female (Statistics New Zealand, 2018), with an estimated age median of 41.7 years (Statistics New Zealand, 2018). In terms of employment, Ashburton's residents are employed in a number of categories, the most prominent of which is managers (24%), followed by professionals (12%), technicians and trade workers (12%), community and personal service workers (6%), clerical and administrative workers (9%), sales workers (8%), machinery operators and drivers (8%), and laborers (21%) (Statistics New Zealand, 2018). Additionally, an estimated 71.5% of Ashburton's residents aged 15 years and over have some type of formal qualification, from which 10.6% have completed at least a bachelor's degree (Statistics New Zealand, 2018).

The ethnic makeup of Ashburton is predominantly comprised of residents who consider themselves of European descent (88.3%), followed by Māori (7.3%), Pasifika (3.4%), Asian (3.9%), Middle Eastern, Latin American or African (0.9%), and other (1.7%) (Statistics New Zealand, 2018). Like with the town of Glenorchy described in this chapter, Ashburton has an ethnic distribution that surpasses 100% due to Statistics New Zealand's coding system, which allows residents to input more than one ethnic background (Statistics New Zealand, 2018).

About the Participants

Data collection in Ashburton took place between April 16 and 26, 2018, during which time 25 respondents agreed to take part in the study. All the respondents who participated in this study either resided or were employed in Ashburton. In terms of employment, participants reported a wide range of occupations, which are summarized in Fig. 5.2. Considering the variety of income-generating sources reported by the respondents, it is anticipated that their observations will provide a clear and representative indication of the state of opinion regarding water issues in Ashburton, beyond any one industry.

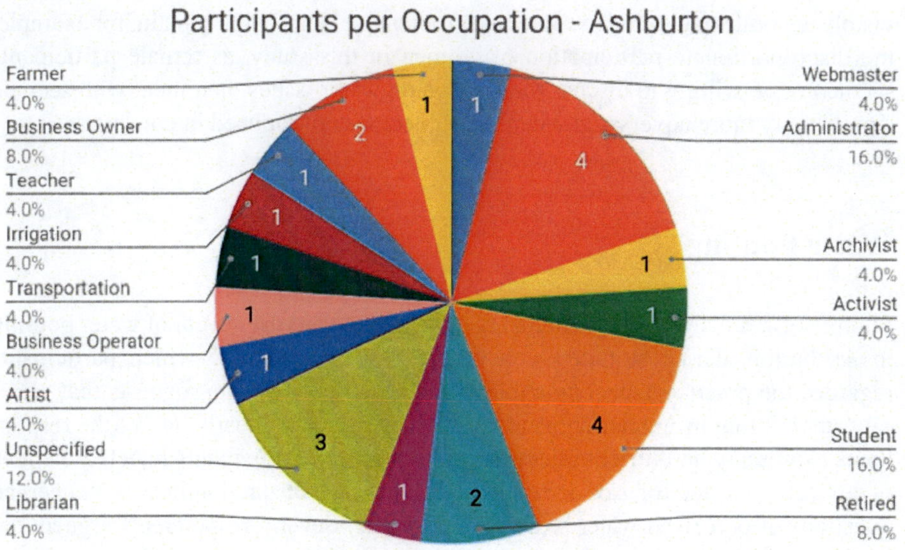

Fig. 5.2 Participants per occupation in Ashburton

Additionally, the age distribution of the respondents was relatively equal: 20% (5) of the respondents were between 18–30, 24% (6) between 30–40, 28% (7) between 40–50, 24% (6) between 50–60, and 4% (1) were 60 or over. Notwithstanding the low participation rate in the 60 and over age bracket, this study reflects a wide range of views across most age groups.

While the participant distribution related to occupation and age was relatively equal, there was a visible imbalance with regard to the respondents' gender: from the 25 participants that agreed to take part in this study, 16 (64%) were female and 9 (36%) were male. In this regard, the respondents' gender distribution does not reflect Ashburton's, which as observed above is significantly more balanced.

In terms of ethnicity and immigration status, the majority of participants reported being of European descent (17), which as observed above, mirrors the overall ethnic makeup of Ashburton. Additionally, three respondents were of Māori descent, four were first-generation immigrants, and one said they considered themselves both Māori and European. The participants who said were immigrants reported having migrated from the UK, Hong Kong, and Iran. Based on the above distribution, this analysis does not adequately represent Māori values and interests related to freshwater in Ashburton, nor on engagement by Māori in hydropolitical conflicts in this locality.

At this stage, it is important to make note of some difficulties experienced during the data collection process. Locals were normally amiable and approachable and seemed generally amenable to have a casual conversation. This positive predisposition however often changed when the topic of water was introduced in the conversation, driving several potential participants to abstain from participating in the study. This took place on more than one occasion, and in several instances, locals became visibly uncomfortable, and in some cases, defensive. This helps explain, for example, the disproportionate participation of women in this study, as female participants seemed more willing to discuss their views on water issues than men, who seemed significantly more adversarial when the subject was introduced in conversation.

Water Bottling

Figure 5.3 summarizes the participants' attitudes toward the advent of water bottling in Ashburton. These attitudes were shaped by the extent to which participants regarded the practice as incompatible in their district. The data suggests that views of water bottling in Ashburton were negative overall, with nearly half of the respondents expressing an outright rejection of the industry. Participants rejected the idea of extracting water for sale for several reasons, which predominantly concerned questions of perceived water scarcity and quality. Several respondents argued that freshwater in the Ashburton District had been over-exploited for profit by a number of different commercial ventures that, by the time the data was collected, were already regarded by the participants as being the main sources of water contamination in the district. One respondent, for example, reported accounts where the ADC discouraged mothers from using tap water to nourish their babies, on the grounds that if they did, their children would develop what the respondent called 'blue baby syndrome'. Concerns over infant health had been elaborated previously by a health official in Canterbury, who argued that nitrates found in water could prove fatal to newborns (Radio New Zealand, 2017). Similar worries have since been articulated by other members of the scientific community (Radio New Zealand, 2021). The above respondent also expressed anger that ratepayers in the district had access to poor-quality water, while greater commercial operations benefited from the best-quality water for what they considered to be a token sum. The water bottling operation was seen as one such enterprise, and as one that did not adequately reveal its real economic benefits to the community, and the extent to which it may have been compatible with the community's interests. In this regard, one respondent said:

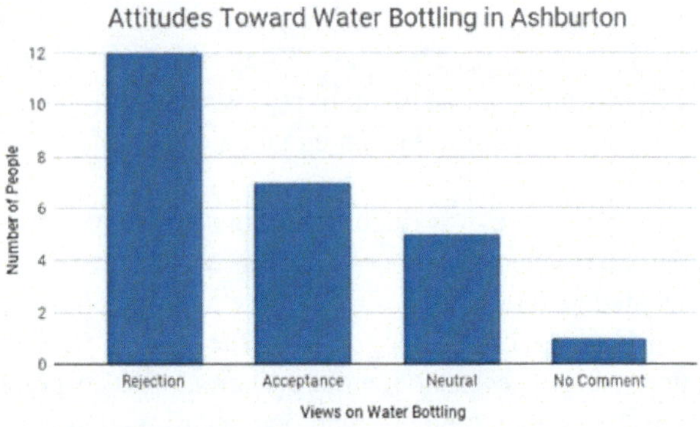

Fig. 5.3 Views on water bottling in Ashburton

> The bottling project was heavily resisted by the community because there was never any disclosure of the economic benefits of the venture, whether the plant would bottle the water on site and ship it later. Later we found out that the water would be sent in bladders in railway carts, so the economic benefits the company was purporting were not accurate. (Respondent 1)

Another participant regarded water bottling with suspicion because, in their view, it created a scenario where water was no longer usable by the local community, due to it being shipped to overseas markets, unlike other ventures in the district. This relates to the high exclusion potential inherent in a water bottling operation, as elaborated in Chap. 3. In this regard, and when asked whether freshwater in Ashburton should be commercialized, they said:

> Absolutely not. Although as I am saying that, I don't want to be a freak about it. The bottling plant for example, when it [water] is taken, it is gone. With farmers there is at least some recharge. (Respondent 6)

A second participant also regarded water bottling as a potential cause of inadequacies. To explain their disapproval for bottling, they said the following:

> It doesn't seem fair, because people have an eye for a profit. You can see places like South Africa where they need to bring water in to make up for shortages, same with Christchurch. One day we decide to sell, and another we end up with nothing. A lot of people have short sighted views, and do not consider leaving their farms to their kids, and focus on short term gain, so there is no stewardship. With such a mentality it is unsurprising that people go from one place to the other depleting as they go. (Respondent 19)

The participant referred to the water crisis experienced in Cape Town, South Africa in early 2018, where the city's water reservoirs fell to dangerously low levels. The incident prompted officials to stop water provision in some areas and ration its water resources across the city (Global Citizen, 2020).

The inevitable exclusion inherent in water bottling motivated additional concerns by other participants, who argued that water authorities should stress the question of intergenerational justice. On this note, one participant argued that:

> [residents] should be able to conserve [water], and not put it to commercial [use]. It should be there for the public for many generations, not just because it is profitable. It is there for generations to come. (Respondent 9)

Negative attitudes toward water bottling were further motivated by non-commercial arguments over natural resource management in the district, and in New Zealand as a whole. Three respondents alluded to Indigenous principles of environmental care and oneness with nature as rationale for disagreeing with the bottling industry, or any practice that commodifies freshwater. On this issue, one respondent expressed that *our Indigenous people value water highly, and we have a Treaty that we need to honor*, when referring to the importance of upholding environmental well-being in New Zealand. When explaining why water is important to him, a second respondent said:

> Because it defines who we are. I am not Māori, but water defines who I am. My family moved from England in 1967. They travelled by water. If you talk to Māori people, they'll tell you where their mountain and river are. Their waka (canoes) came to this land by water. All their stories and legends come back to water. This is where I come from. I have nowhere else to go. This is my home, so I have adopted many of those Māori beliefs into my life. It is the definition of who we are. (Respondent 23)

A respondent explained their rejection toward water bottling in their district by arguing that the industry was incompatible with Ashburton's environmental conditions. On this note, they said the following:

> [Bottling] to me speaks of how councils make decisions: Learn where it is best to bottle water, don't just do it wherever. Sure, we need lime, but don't take it from all mountains. It is about sharing, and not about degrading certain areas. (Respondent 6)

A second respondent evoked the same views over the environmental compatibility of water bottling in Ashburton. To sustain their comments, they said the following:

> From a historical point of view [Ashburton] is very dry. It is very windy. You could only grow wheat crops and sheep. People usually think this area is lush and green, but don't realize it is not natural. That water comes from somewhere else. (Respondent 4)

Several other respondents stressed the connection between unhindered access to clean water and their ability to exercise their local and national identities. In this regard, one respondent argued that their ability to be a New Zealander is directly related to their unrestricted access to nature, and to recreational activities such as hiking, fishing, and swimming, all of which are highly dependent on good water quality and quantity. Two other respondents coincided with the recreational aspect of their water use, quoting memories from their childhood of swimming or interacting with wildlife around rivers and lakes. In particular, one participant pointed out that in order for residents to engage in these types of activities, they need to go to the top-most part of the district's rivers, where freshwater remains mostly untouched by human activities and is therefore sufficiently clean. In this regard, they said:

> If you want to go swimming now, you have to go right up to the hills [Southern Alps] now, to the source of the water. It is not the best part of the river because it is the coldest part. (Respondent 20)

These respondents also concurred on a common dislike toward water bottling and toward the commercialization of water in general.

The data also documents some accounts in favor of water bottling, and expressions of conditional support for the industry. However, these accounts do not nullify the number of voices that expressed skepticism toward water bottling. In this regard, two respondents expressed open support for the industry, while five conveyed conditional support. The attitudes in favor of water bottling in Ashburton are perhaps best related by one respondent, who pointed out that the activity is likely to extract far less water than enterprises like irrigation and farming. This respondent also stressed the significant volume of freshwater in the district's aquifers, and in the three riverways that cross the Ashburton area: the Rakaia, Rangitata, and Ashburton rivers, all of which are fed by glaciers and snowmelt from the Southern Alps. To substantiate their arguments in favor of the water bottling operation in Ashburton, they said the following:

> [The] land came with a consent to take water. And it was I think 48 L per second, which in the greater scheme of things here, is not a lot. One of the conditions of the consent was that the groundwater needed to be replenished from a river, at a rate of 64 L per second. River water is alpine water, so ice and snow melt that comes down. So, they're taking that water and pumping it into the aquifers. No one took any notice of that clause. They focused on the part that said that the venture would pump 48 L [of water] per second... and [that it would have sold] it to an overseas company that is going to bottle and sell it: Who cares? (Respondent 21)

In this respondent's view, water abundance helps to justify the extraction and commercialization of freshwater in both the district and in a country that has *more water than it knows what to do with.* Alongside these reasons, positive attitudes toward water bottling were based on the establishment of sound environmental impact assessments, regulations, and the locals' ability to benefit from water irrespective of industry.

Water Chlorination

Unlike water bottling, water chlorination seems to inspire mixed attitudes in the participants. There is a significant number of respondents that neither accepted nor rejected the policy: out of the 25 participants in this study, 15 did not adopt any perceivable attitude toward water chlorination. Conversely, one-fifth of the respondents expressed an explicit dislike or disapproval for water chlorination in their district, with only one espousing positive attitudes toward the latter. Four additional respondents did not make comments on the matter (Fig. 5.4).

The accounts documented in this study suggest that respondents may have internalized the implementation of water chlorination to the point where adopting positive or negative views over the policy was precluded by the real need to securitize the district's water supplies. One potential cause for this lack of clear inclination may be attributed to the principal sources of water quality degradation in the district,

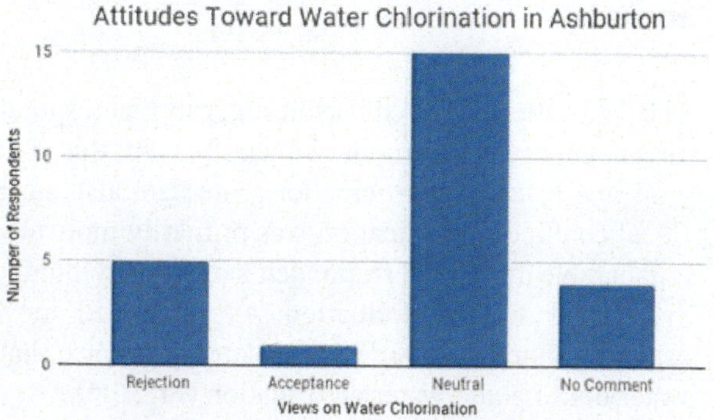

Fig. 5.4 Views on water chlorination

which have been linked to intensive dairy farming, the main source of economic well-being in Ashburton. If environmental decay is seen as an inevitable and irreversible side-effect of the main generator of wealth in the district, one could expect that respondents would be disinclined to espouse strong attitudes toward water securitization strategies such as chlorination. This assertion seems to be supported by the data. From the 15 respondents who expressed neutral views over water chlorination, eight reported dissatisfaction with the quality of their water, which suggests that respondents may regard chlorination as a necessary and unquestionable mechanism to secure their water supplies. It did not seem that respondents perceived water chlorination as being incompatible in their district.

Those participants who rejected water chlorination argued that the policy had had negative impacts on their personal health and on their ability to address their own food requirements. This rejection also highlighted asymmetries in water access and use in the district. One respondent, for example, said the following:

> People developed skin conditions from using chlorinated water and could not grow their own crops as a result of that. This is substantial given that many people grow their own food in the district. Why should I take a shower and itch, use bad water, when they can get all the best water for free? (Respondent 1)

The data points to a trend worth noting in those respondents who reported strong views against water chlorination. The five respondents who expressed a strong dislike toward water chlorination also reported some type of anguish with regard to at least one more water-related policy or practice. For example, two respondents expressed disapproval toward chlorinating Ashburton's water supplies in conjunction with negative attitudes toward water bottling. Two other respondents reported being severely disillusioned with the decision-making process behind any water-related practice or policy, echoing concerns related to the favoring of commercial enterprises over other less-commercial interests. The respondents in question disapproved of water chlorination but regarded the policy as one in a handful of water mismanagement cases in the district. The fifth participant documented having supported strikes against water treatment in the district; this treatment included both water chlorination and fluoridation.

Results

The data collected in Ashburton suggests that respondents espoused a strong willingness to engage in hydropolitical conflict, but this conflict intentionality was coupled with low levels of participation. The data also suggests that the risk of hydropolitical conflict intensification was primarily motivated by water bottling, although a comparable number of respondents reported willingness to engage in intense conflict events over water chlorination. As mentioned above, however, these respondents espoused more than one water-related grievance that, in some cases, included both water bottling and water chlorination (Fig. 5.5).

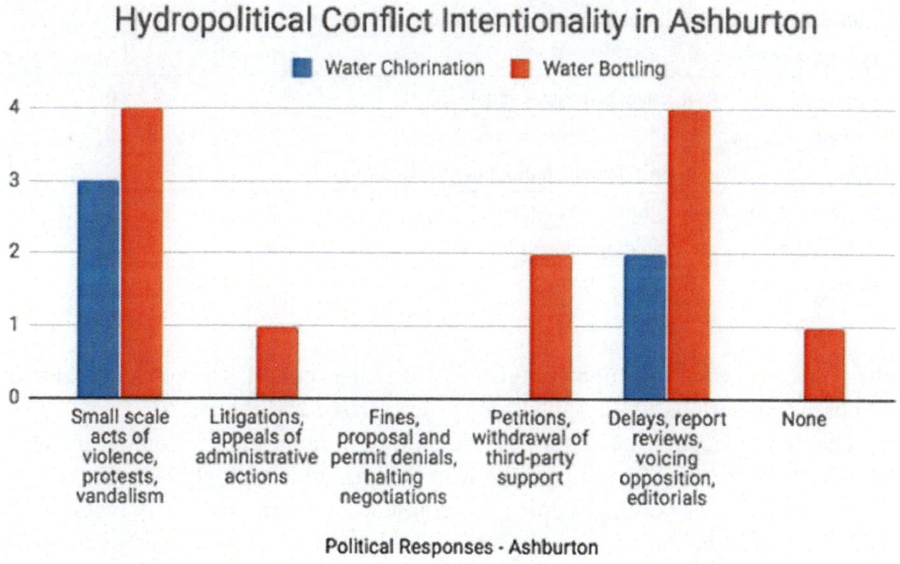

Fig. 5.5 Hydropolitical conflict intentionality in Ashburton

In general, water bottling and water chlorination motivated respondents to espouse hydropolitical conflict intentionality at different levels. For example, 11 out of 25 participants reported being willing to engage in conflict over water bottling, while only five expressed similar inclinations over water chlorination. From the individuals that expressed a desire to organize politically against water bottling, four were inclined to do so by starting or participating in protests, while a fifth participant reported being willing to support litigation procedures against the enterprise. In contrast, three out of the five respondents were willing to engage politically against water chlorination through protesting, while none were willing to engage in legal challenges against the policy.

Along the less-intense section of the CCI Scale, two respondents argued they would prefer petitioning to voice their disapproval of water bottling, while four regarded voicing opposition through their elected officials as the best option to express their discontent for the same enterprise. In contrast, water chlorination only motivated two instances where participants were willing to seek access to their elected officials to voice disapproval for the policy. These responses are summarized in the graph above.

In terms of gender, the data indicates that there was a clear preference by female respondents to engage in intense hydropolitical conflicts: four respondents reported either being willing to be or having been involved in protests against water bottling and/or water chlorination, while one additional respondent expressed a will to challenge these activities through the courts. In stark contrast, no male respondent reported wanting to engage in conflict action at either one of these levels. These dynamics changed to some degree in the lower levels of the CCI Scale, where both female and male respondents reported comparable conflict intentionalities, although

Table 5.1 Gender distribution of hydropolitical conflict intentionality in Ashburton

Type of Action	Female	Male
Small scale acts of violence, protests, vandalism	4	0
Litigations, appeals of administrative actions	1	0
Fines, proposal and permit denials, halting negotiations	0	0
Petitions, withdrawal of third-party support	2	1
Delays, report reviews, voicing opposition, editorials	5	3
None	4	5

these numbers were still slightly higher for the female cohort. The gender distribution of conflict intentionality in Ashburton is summarized in Table 5.1.

The data indicates that water bottling motivated more intense conflict intentionality in Ashburton when compared to water chlorination. A highly extractive enterprise such as water bottling seemed and continues to seem, inherently incompatible with the water dynamics of districts such as Ashburton, where water access and policy are attached to the development of the dairy and irrigation industries. During the data collection, respondents were quick to attribute the root causes of water degradation to these industries, while at the same time remaining cognizant of their role in the generation and sustainability of wealth in Ashburton, and in the Canterbury region as a whole. When asked about their views on water-related decision-making in the Ashburton District, one respondent said the following:

> It is governing to a small percentage of the population, the farmers, which is fine, we need them, and dairy. But farming is becoming too intensive. It is starting to bite them in the ass. Eventually these rivers are going to be unusable for anybody, so a small percentage of the population having the right to affect the rest of the country like that is unfair. (Respondent 20)

This duality is important to consider when understanding why conflict emerged and intensified over water bottling, but not over dairy farming and irrigation. Based on the accounts documented in this study, one can roughly surmise that a balance of sorts existed prior to the prospective introduction of water bottling in Ashburton. Residents and the more traditional industries seemed to coexist with an unwritten understanding that industry-driven water degradation could be somewhat tolerated by the community if it was met with wealth redistribution, employment generation, and the passing of water securitization and treatment strategies, all of which would arguably mitigate the deterioration of the district's water. This is not to say that the dairy and irrigation industries were not perceived as sources of water problems by the respondents; it means that efforts were put in place to mitigate the adverse environmental and social impacts of these enterprises. In this regard, the data indicates that water bottling was regarded as an operation that had a significant potential to worsen the residents' domestic water use, as well as posing challenges to the water access parameters of the more traditional industries in the district, all the while offering remedial means that were regarded as inadequate by some respondents. These means included diverting some of the Ashburton River's flow and using that water to recharge the aquifer system from which the bottling plant would have extracted its supplies, thus mitigating the quantitative impacts of the operation (Radio New Zealand, 2016). Some respondents argued that this measure would have proven costly and that taxpayers would have been the ultimate bearers of the costs of replenishing the aquifer (Radio New Zealand, 2016).

The mitigation strategies conceived by the dairy and irrigation industries were not replicable by the water bottling operation. Unlike the dairy and irrigation industries, which the respondents regarded as producing significant economic opportunities in the district, a water bottling operation would have been considered to benefit from relatively inexpensive access to significant volumes of high-quality water (New Zealand Herald, 2016), all the while inspiring doubts regarding its real employment generation potential in the district (New Zealand Herald, 2016). In this instance, the coexistence that the dairy and irrigation industries were successful in fostering with the community seemed significantly less feasible for the water bottling operation; the latter motivated community members and local industries to coalesce in an effort to thwart the installation of an enterprise that both cohorts seemed to regard as incompatible in their district.

Discussion

Bearing in mind the above findings, one can argue that the data collected in Ashburton supports this study's first argument: the introduction of water commercialization practices seemed to influence the emergence of hydropolitical conflict intentionality if it was seen as the source of economic, environmental and social incompatibilities in the local community. In this regard, more respondents espoused negative views over water bottling than over water chlorination, although the number of respondents relative to the total amount of participants was fairly low in both categories. Furthermore, conflict intentionality over bottling was concentrated at both high- and low-intensity levels, namely at the Protest and Voicing Opposition scales.

This study finds evidence suggesting that water bottling in Ashburton sparked a strong political response because it was in tension with the interests of major economic enterprises in the district, namely actors in the dairy and irrigation industries. In this regard, the main political organizer of the 2016 Ashburton protests reported that the movement against water bottling was initially supported by representatives of the farming and irrigation industries, who later withdrew from the movement when their industries' water use was put into question. Those same economic interests went unchallenged in the chlorination of Ashburton's freshwater. To some extent, one can argue that water chlorination achieved the dual goal of securitizing the district's freshwater while appearing to fix, in some measure, the negative environmental impacts of the dairy and irrigation industries. Thus, the chlorination of Ashburton's freshwater seemed to have mitigated the incompatibilities caused by other industries in the periphery, which helps explain why respondents seemed less likely to espouse intense hydropolitical conflict intentionality over the policy. In contrast, the respondents seemed to regard the water bottling operation as an enterprise with low social and economic remediation potential, and unlikely to assuage its environmental footprint in the district. In terms of economic, social, and environmental incompatibilities, water bottling was regarded as the enterprise that was least suitable for Ashburton's local reality.

This study also argued that hydropolitical conflict risk is likely to intensify when individuals regard the approval and appeals process behind a water commercialization activity as untrustworthy. The data supports this assertion: respondents displayed significant skepticism toward the approval process of water bottling, which in turn translated into both intense hydropolitical conflict intentionality and similarly intense conflict events such as the 2016 protests. There is indeed evidence that proper consultation with the community was not started by the ADC with regard to the sale of Lot 9. For example, concerns that local Māori leadership was not adequately engaged were an issue highlighted in ADC records:

> Council's Chief Executive Officer and Mayor have received correspondence from Ngāi Tahu and Arowhenua Rūnanga about their concerns with the Lot 9 sale, stock water closures and the impact on Mahinga Kai. The main issue highlighted is one of engagement with local iwi for water matters in general. (Ashburton District Council, 2016, p. 30)

Files submitted by the Bung the Bore movement echoed the above concerns regarding community feedback and input, but also stressed the ADC's alleged role in not including the community as an Affected Person as per the RMA framework:

> We are at a loss as to how the Ashburton District Council came to the conclusion that Mr. and Mrs. 'C' and Silver Fern Farms would be the only parties affected by the Lot 9 take. Subjecting ratepayers to the expense of lowering bores for affected parties, paying their legal fees, and assuring them a clean and continuing water source should have been a community decision, just as the giving away of 40 billion liters should have been. (Ashburton District Council, 2016, p. 39)

The Affected Persons to which the group alluded above were also the ones included in the resource consent application by the ADC before ECan, the water authority responsible for granting such consents. This application preceded the sale of Lot 9 and was submitted in 2011. In their application, the ADC reported the following:

> [F]ive bores were shown to be potentially affected. One bore is owned by ADC and does not need to be considered, three are owned by Silver Fern Farms, and one is owned by Mr and Mrs JL and J Cockburn. Written approvals have been obtained from all potentially affected parties, being Silver Fern Farms and Mr and Mrs Cockburn. (Environment Canterbury, 2011, p. 2)

Community groups that engaged in intense conflict over the sale of Lot 9 for bottling later organized an effort to have the property's water consent removed, threatening to go to the courts over the issue:

> We assure you of our intent to take this consent and it's [sic] process to Judicial Review should common sense not prevail. In doing so, the people of Ashburton, and indeed New Zealand, would be asked to fund a challenge to save what is already 'ours'. Should there be no decision made to abandon this consent by July 30th [2016], we will be left with no other choice than to instigate legal proceedings. (Ashburton District Council, 2016, p. 41)

The above statements demonstrate the extent to which some members of the community felt the approval process for the water bottling operation was flawed. They also provide an indication as to why some respondents were likely to exhibit intense hydropolitical conflict intentionality. In contrast, respondents gave little indication that the approval and appeals process behind the chlorination of Ashburton's freshwater was in any way flawed, or untrustworthy. This is not to say that chlorination inspired overly positive attitudes, as there are accounts in the data that suggests that some respondents found chlorinated water problematic because of its look, taste, and the health problems attributed to chlorine exposure. Nevertheless, accounts of this nature seemed to have been overshadowed by the legitimacy behind the policy, and by the urgent need to securitize the district's freshwater.

The case of Ashburton raises crucial points in relation to perceptions of water issues and how those perceptions influenced some members of the community to espouse some type of hydropolitical conflict intentionality. In addition, the trust that respondents invested in the approval and appeals process behind any given water commercialization practice seems to dictate the extent to which they are likely to intensify said intentionality, and potential engagement. Both water bottling and water chlorination inspired some type of discontent in the community, but this discontent translated into conflict primarily in relation to bottling. Water bottling motivated perceptions of water scarcity in Ashburton, but these views were exacerbated by a seemingly flawed approval and appeals process, leading respondents to circumvent the mechanisms inherent in the water authority system, and to seek extra-institutional ways to sway decision-making in their favor. These extra-institutional means resulted in the organization of political protests.

With regard to chlorination, considering that most respondents perceived water contamination as being a result of the direct and indirect activities of the dairy and irrigation industries and that addressing the root causes of contamination would inevitably require a de-intensification of these industries (and therefore a potential reduction in their economic outputs), it is likely that respondents were discouraged from rejecting any water securitization strategy. Therefore, it can be argued that water chlorination facilitated an improvement in water access and use in the district. There is little indication in the data that suggests that respondents distrusted the approval process behind water chlorination, which also helps to explain why they were less likely to espouse intense hydropolitical conflict intentionality over this policy.

References

Ashburton District Council. (2016). *Agenda for council meeting on 28 July 2016*. Retrieved May 26, 2019, from https://www.ashburtondc.govt.nz/our-council/meetings/agendas-and-minutes/AgendasAndMinutesDocs/CouncilAgenda28July2016.pdf

Environment Canterbury. (2011). *Officer's report—CRC111706 & 111707*. Retrieved June 1, 2019, from https://api.ecan.govt.nz/TrimPublicAPI/documents/download/2562520

Global Citizen. (2020). *Cape town is about to run out of water*. Retrieved February 6, 2022, from https://www.globalcitizen.org/en/content/cape-town-water-crisis-day-zero/

New Zealand Herald. (2016). *Ashburton council reneges on controversial water bottling deal—NZ Herald*. Retrieved August 24, 2017, from http://www.nzherald.co.nz/business/news/article.cfm?c_id=3&objectid=11672314

Radio New Zealand. (2016). *Ashburton ratepayers to cop aquifer refill bill*. Retrieved August 24, 2017, from http://www.radionz.co.nz/news/regional/301165/would-be-ashburton-water-bottling-company-named

Radio New Zealand. (2017). *Concerns raised over nitrates' effects on babies*. Retrieved April 10, 2019, from https://www.radionz.co.nz/news/national/341701/concerns-raised-over-nitrates-effects-on-babies

Radio New Zealand. (2021). *Drinking water nitrate limit 11 times higher than it should be—health expert*. Retrieved January 27, 2022, from https://www.rnz.co.nz/news/national/436088/drinking-water-nitrate-limit-11-times-higher-than-it-should-be-health-expert

Statistics New Zealand. (2018). *2013 Census QuickStats about a place: Ashburton District*. Retrieved March 7, 2018, from http://archive.stats.govt.nz/Census/2013-census/profile-and-summary-reports/quickstats-about-a-place.aspx?request_value=14909&tabname=Business

Chapter 6
Analysis

Abstract This chapter provides a tripartite analysis of the conflict-inducing potential of water bottling and water chlorination in Ashburton and Glenorchy. It concludes that the emergence of hydropolitical conflicts in both locations was determined by the economic, environmental, and social compatibilities of water bottling and water chlorination, and that the risk of conflict intensification was determined by the extent to which residents trusted the approval process behind each operation.

Introduction

In the context of water abundance in New Zealand in general, and through the application of what was proposed as the Coleridge Dilemma in particular, this study advances two central arguments. First, that water commercialization generates hydropolitical conflict intentionality when the salience of water scarcity or quality is raised. The data collected in Ashburton and Glenorchy supports this argument. Second, that the risk of hydropolitical conflict intensifies in accordance with the extent to which communities trust the approval and appeals processes behind a commercial operation. Water bottling and water chlorination, operations that fall within the framework of the proposed definition of water commercialization presented in Chap. 3, exerted an influence on the emergence of hydropolitical conflict. However, the data also indicates that not all conflict intentionalities intensified to the top-most levels of the CCI Scale, leaving a significant amount of accounts and events dispersed across the other intensity levels of the Scale. The data suggests that the emergence of conflict intent was motivated by concerns over incompatible water activities in each district, but that its intensification appears to have been motivated by an erosion of trust in the approval and appeals processes behind the scrutinized water bottling and water chlorination projects.

This chapter will explain first, how water bottling and water chlorination motivated the emergence of hydropolitical conflict intentionality in Ashburton and Glenorchy; secondly, how respondents sought to address water bottling and chlorination issues through their existing water authorities. In this regard, this section will highlight the

A. E. Suazo, *The Risk of Water Conflicts in Aotearoa-New Zealand*,
https://doi.org/10.1007/978-3-031-06660-3_6

question of trust in the approval and appeals processes supporting the above practices, and how this trust played a role in respondents' engagement in low-intensity, intra-institutional conflict. And thirdly, this chapter will explain how untrustworthy or perceivably flawed approval and appeals processes motivated respondents to circumvent their water authority's conflict resolution mechanisms, and instead increased the risk of hydropolitical conflict intensification over water bottling and water chlorination.

The Emergence, Containment, and Intensification of Hydropolitical Conflict Intentionality

Figure 6.1 maps the trajectory followed by the respondents in Ashburton and Glenorchy as they contemplated their potential engagement in low- and high-intensity conflicts over water bottling and water chlorination. At this stage, one must take note of two important considerations. First, the diagram below and the observations in this chapter are derived from a relatively small respondent sample, which

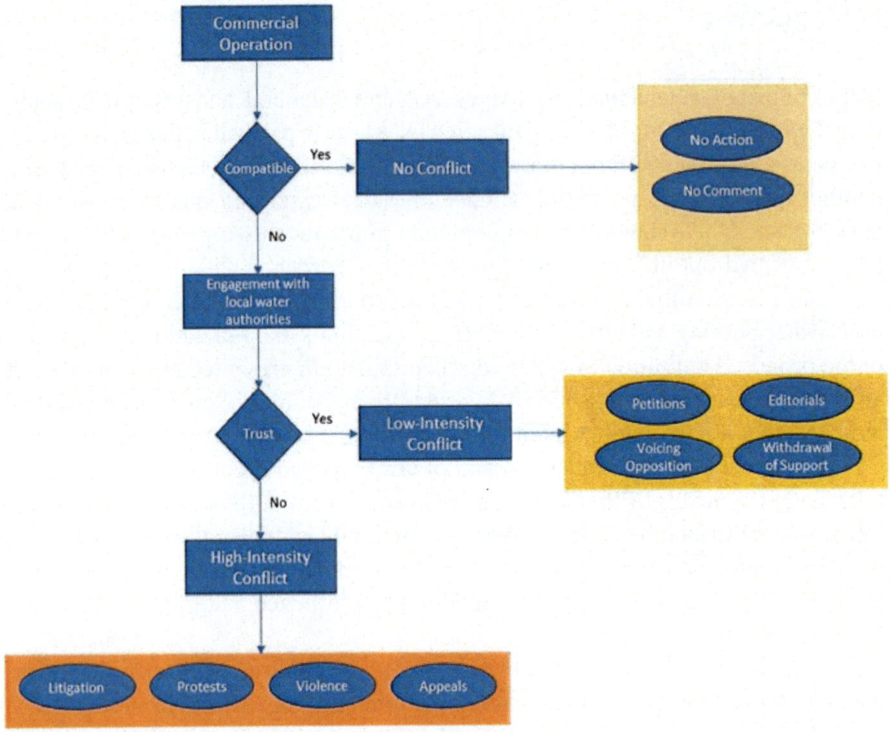

Fig. 6.1 Flowchart depicting pathways toward potential hydropolitical conflict intensification

cannot be construed as reflecting the overall views, conflict intentionalities, and political engagement of the wider communities in Ashburton and Glenorchy. Notwithstanding this limitation, the observations gathered from these samples provide important insights into the general thought-process of individuals conflicted by the introduction of a commercial enterprise they regard as antagonistic. In addition to this, the data demonstrates the level of common ground the respondents were capable of achieving irrespective of their demographic traits.

And second, the above pathway was developed based on the respondents' conflict intentionalities as expressed by them during the interview process and does not adequately reflect the respondents' actual hydropolitical conflict participation. Because the arguments presented in this study relate to conflict intentionality and risk intensification, and not to the materialization of conflict, it is considered that the above pathway provides an appropriate account of the steps that may antecede the crystallization of hydropolitical conflicts over activities such as water bottling and water chlorination.

There are three discernible processes that respondents seemed to follow in both locations: one where conflict intentionalities were motivated by concerns over water bottling and water chlorination; a second one where such conflict intentionalities were effectively contained by the water authority system in each district; and a third where the risk of conflict intensification increased. The causal path that led respondents to any of these outcomes seemed to be decided by two crucial pivot points, the first of which is the respondents' perception that water bottling and water chlorination could cause water issues in their districts, in line with this study's first argument.

First Pivot Point: The Identification of Water Incompatibilities

The data indicates that water bottling and water chlorination inspired different attitudes by the respondents. These attitudes were shaped by how they identified each water practice as causing incompatibilities locally. These attitudes are summarized in Figs. 6.2 and 6.3. In Ashburton, water bottling motivated predominantly negative attitudes, but respondents expressed inconclusive opinions regarding the chlorination of their freshwater supplies, as evident in the number of neutral views or no-comment statements over the policy. In Glenorchy, respondents espoused predominantly neutral views over bottling, although a slightly higher number reported holding positive rather than negative attitudes toward the industry. As in the Ashburton water bottling project, Glenorchy's respondents espoused overwhelmingly negative attitudes toward water chlorination where, as explained in Chap. 4, locals did not seem to regard the policy as having any real application in their immediate surroundings, an issue not so readily seen in Ashburton, where respondents were much more aware of the risk of accessing low-quality freshwater, and of the need to treat it.

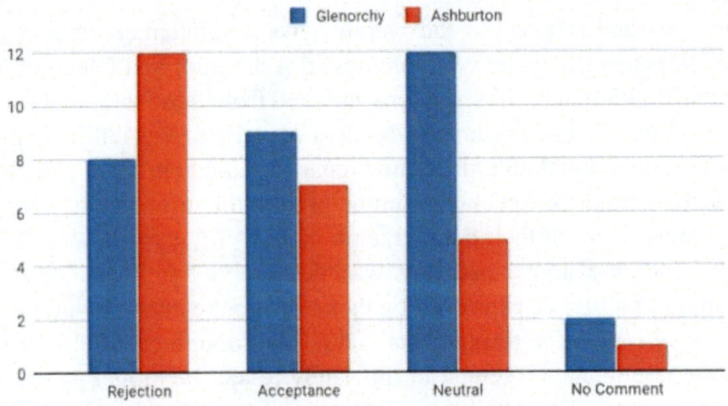

Fig. 6.2 Attitudes toward water bottling in Ashburton and Glenorchy

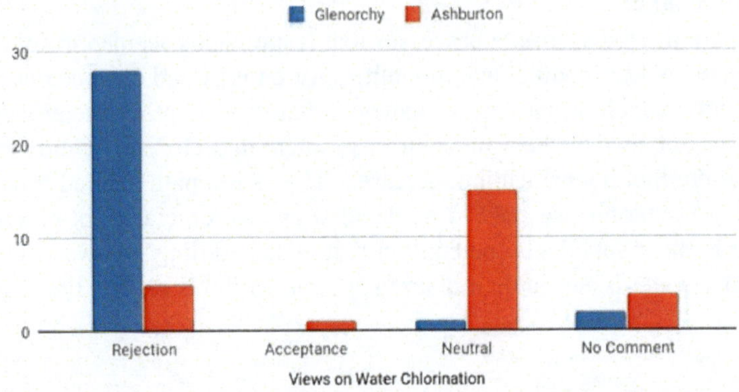

Fig. 6.3 Attitudes toward water chlorination in Ashburton and Glenorchy

Based on the data presented above, one can surmise that respondents in Ashburton and Glenorchy determined that the likelihood of water-based incompatibilities was more pronounced in the face of water bottling and water chlorination, respectively.

Second Pivot Point: The Containment and Intensification of Hydropolitical Conflict

Containment

Once respondents perceived that there could be issues caused by water bottling or water chlorination, they were confronted with a second determination. This point relates to the question of trust in the legitimacy of the approval and appeals processes supporting water bottling and water chlorination within their local councils. Those respondents who expressed no doubts over the above practices effectively engaged in intra-institutional exchanges with their water authorities, to ventilate any given water-related grievance and/or incompatibility. In this case, conflict intentionality remained at the lowest possible levels of the CCI Scale.

As per Chap. 2, the RMA regime establishes the criteria applied by water authorities to decide upon issues of freshwater access and use. It also considers additional processes for the effective resolution of disputes, should any arise. Considering the levels within the CCI Scale, hydropolitical conflicts that one may consider of low intensity are effectively ventilated and resolved through councils' RMA-enabled processes.

One can argue that the success of the above processes hinges on whether applicants and submitters deem the entire regime as trustworthy or not. In this regard, the data suggests that respondents' inclination to engage in low- or high-intensity hydropolitical conflict was based on the extent to which they trusted the approval and appeals processes behind water bottling and water chlorination in their districts. To this effect, Table 6.1 displays a distribution of all accounts reporting negative attitudes toward water bottling and chlorination and distributes them in relation to the most intense hydropolitical conflict event in which each respondent would be willing to engage.

When analyzing the respondents' conflict engagement potential, the data indicates that intentionality remained low when they trusted their local authorities' decision-making and management of each practice. This, in turn, translated into hydropolitical conflict intentionality becoming contained, both at the lowest levels of the CCI Scale, and at levels where water-related decisions are handled locally. For instance, the approval process behind water bottling in Glenorchy was not regarded as problematic

Table 6.1 Hydropolitical conflict intentionality in Ashburton and Glenorchy

	Glenorchy		Ashburton	
	Water chlorination	Water bottling	Water chlorination	Water bottling
Small scale acts of violence, protests, vandalism	0	0	3	4
Litigations, appeals	11	1	0	1
Fines, proposal and permit denials, halting negotiations	0	0	0	0
Petitions, withdrawal of third-party support	2	0	0	2
Delays, report reviews, voicing opposition, editorials	10	7	2	4
None	8	0	0	1

because the purchase of the operation's potential site (and its accompanying water rights) came as a result of a private transaction between the latter's previous owner and the bottling company, a process respondents did not seem to consider worthy of disruption nor skepticism. In addition to this, and as was explained in Chap. 4, the consent approval process, as administered by the ORC, was not judged by the respondents to be inadequate or flawed. In terms of conflict intentionality, those respondents who disagreed with water bottling (eight out of 31) expressed little inclination toward intense political organization and engagement, which also helps to explain why conflict intentionality over bottling in Glenorchy is concentrated in the 'Voicing Opposition' level of the CCI Scale, a space where conflicts over water are ventilated locally through the institutions and processes laid out by the RMA framework. In other words, the risk of intense hydropolitical conflict over water bottling in Glenorchy is argued to be low.

Similarly, the chlorination of Ashburton's water supplies did not seem to be regarded as a policy that was approved under dubious conditions. In fact, chlorination was perceived as part of a wider water securitization strategy that also featured updates in water infrastructure and other forms of treatment, including fluoridation. No accounts documented in this study suggest that respondents deemed the decision-making process behind chlorination as tainted, or as illegitimate. With respect to the risk of conflict intensification over water chlorination, five out of 25 respondents expressed disapproval over the policy, but potential conflict engagement within this cohort was very low, with only three participants willing to engage in intense conflict, and a comparable number of respondents (two) expressing a will to voice their discontent with their local elected officials. Like with water bottling in Glenorchy, the risk of intense hydropolitical conflict over water chlorination in Ashburton was considered low.

In the above cases, there is little indication that the decision-making process behind each practice or policy influenced the participants' will to intensify their political engagement beyond interactions with their local water authorities. This seemed to be the case even when respondents regarded water bottling and water chlorination as potential sources of water-based conflict. Although hydropolitical conflict intentionality in Ashburton and Glenorchy did not intensify, it does not mean that conflicts over water bottling and water chlorination were absent: It could mean that the local political infrastructure that decides on water issues, and the processes allowing respondents to express their worries over any given decision, were regarded by the respondents as sufficiently robust and trustworthy to ventilate and assuage any of their concerns intra-institutionally, thus preventing these conflicts from escalating further.

Intensification

It follows from the data, however, that when the integrity of local processes and institutions, and that of their elected officials, was put into question, respondents were more likely to seek measures outside of local councils' jurisdiction to assuage or prevent the prospect of a water incompatibility. Thus, a bypass of sorts seemed to occur, with some respondents disregarding local conflict resolution mechanisms claiming the water authority regime was unlikely to mitigate potential water inadequacies. Instead, these respondents displayed high-intensity, extra-institutional conflict intentionality, presumably to sway the direction of decision-making in their favor. In these cases, the intensification of hydropolitical conflict risk became more likely.

The above expressions of conflict engagement manifested in both the respondents' conflict intentionality, and in the active participation by some of them in conflict episodes motivated by water bottling and water chlorination in Ashburton and Glenorchy, respectively. For example, there were 28 respondents in Glenorchy who expressed negative attitudes toward water chlorination, 23 of whom reported willingness to engage in some type of political and/or judicial conflict. Of those 23 respondents, 11 reported being inclined toward litigation procedures to challenge the chlorination policy, followed by two accounts preferring petitioning and/or removal of third-party support, and 10 arguing that contacting their local officials was the best possible political route to voice their disagreement. In addition to these voices, eight expressed unwillingness to take any action against chlorination, despite their disapproval of it. In Glenorchy, hydropolitical conflict intentionality over chlorination is mainly directed toward litigation, namely the second-most intense hydropolitical conflict level in the CCI Scale. This study also captured evidence suggesting that the risk of escalation to the top-most level of the CCI Scale in Glenorchy was possible.

Similarly, water bottling inspired some actors in Ashburton to intensify their conflict engagement potential beyond local interactions with with councils: from the 11 respondents who expressed being willing to engage in conflict over water bottling, four reported an inclination to protest the enterprise, and one expressed a willingness to go to the courts. A total of five respondents expressed intentions to engage in intense hydropolitical conflict over water bottling. In addition, two respondents reported wanting to start and/or participate in petitioning, while four argued that the best way to counter the water bottling industry would be through their local authorities. The above figures translated into a number of respondents willing to intensify their political engagement beyond their local authorities' purview, although in Ashburton, the number remained low in relation to the total amount of respondents.

In both Ashburton and Glenorchy, high conflict intentionality seemed irrevocably linked to water bottling and water chlorination being authorized by local processes that left little space for respondents to genuinely voice their discontent intra-institutionally. In Ashburton, for example, the approval process for the water bottling operation was regarded with significant skepticism largely because the purchase of the property where the bottling plant was envisioned to operate was part of the Ashburton Business Estate and owned by the ADC, the local district council. Even though the resource consent for this project fell outside of the jurisdiction of the ADC, the above transaction inspired the belief that the potential incompatibilities that had already been identified would have been at least partly facilitated by their local authorities, at a point when respondents were already skeptical over the ADC's ability to ameliorate water quality standards in the district. Also, the project was being considered despite its dubious economic, environmental, and social benefits and ramifications. For instance, Chap. 5 related the concerns expressed by some respondents with regard to the operation's highly debated employment generation potential, which some argued was far less than purported by the operator. They also expressed discontent over the plant's aquifer replenishment strategy, which would have caused part of the Ashburton River flow to be diverted into the aquifer from which water would have been abstracted, infrastructure respondents argued would have been paid for by taxpayers.

From a wider perspective, there are some political dimensions that may have further aggravated the respondents' disapprobation of the water bottling approval and appeals processes. Prior to the proposed water bottling operation in Ashburton, ECan, the authority in charge of approving resource consents for water extraction in the region, was already undergoing an acute crisis of trust, evident in actions taken by New Zealand's central government. The central government's low regard for ECan's ability to govern its freshwater supplies was such that, in 2010, it decided to remove all ECan's elected councilors. It then replaced them with a Council appointed by central government officials (New Zealand Parliament, 2010), with the promise that elections for new councilors would take place no later than 2013 (New Zealand Parliament, 2010). This plan, however, was later amended through the passing of the Environment Canterbury (Transitional Governance Arrangements) Act 2015, which enacted changes to ECan's governance structure. These changes included the overall makeup of the Council itself, which was to be comprised of both elected and centrally appointed officials, amendments that were envisioned to take effect as of the 2016–2019 electoral cycle (Government of New Zealand, 2015). Because the 2016 election was to take place 'on the second Saturday in October 2016' (Government of New Zealand, 2015, p. 4), the approval process for the water bottling operation's resource consent in Ashburton was deliberated upon by an unelected regional council. It is anticipated that the unelected nature of the Council may have contributed to its inability to make decisions that were perceived as trustworthy, and as reflective of citizens' needs. In this regard, the inadequacies that respondents had already seen could materialize because of the water bottling plant, seemed further aggravated by decision-making processes and institutions that respondents did not trust, or whose legitimacy they questioned.

Similarly, respondents in Glenorchy expressed significant disdain over having been inadequately consulted in the QLDC's decision to chlorinate the town's water supplies. As per Chap. 4, respondents documented doubts over the applicability of chlorination in a town where the quality of water had traditionally been high. They further argued that the introduction of chlorine would only cause them problems and that they were being stripped of their status as Affected Persons (as per the RMA framework), and therefore, of their ability to prevent and/or appeal the chlorination decision. This led to a belief that any interaction with the QLDC regime would have proven purposeless, thus justifying the need to conceive of extra-institutional means to facilitate the addressing of chlorine-driven incompatibilities. It is under this light that conflict intentionality over water chlorination in Glenorchy intensified.

Considering the above-described conflict intentionalities, and the subsequent materialization of some hydropolitical conflict events over water bottling and chlorination in Ashburton and Glenorchy respectively, one can surmise that the arguments proposed in this study find support in the data: Water bottling and water chlorination influenced the emergence of hydropolitical conflict intentionalities in Ashburton and Glenorchy. The risk of conflict intensification was further determined by the extent to which respondents regarded the approval and appeals processes behind each practice as trustworthy. As it stands, however, the arguments include two important caveats.

Firstly, the data shows that whenever a commercial water operation inspired an intense hydropolitical conflict intentionality by the respondents, as was the case with water chlorination in Glenorchy and with water bottling in Ashburton, it also provoked a comparable reaction at the bottom-most portion of the CCI Scale, where a significant number of low-scale hydropolitical conflict intentionality accounts is discernible. This indicates that when conflict intentionality escalates, the risk factor does not apply generally across the whole community. Indeed, communities featuring high-intensity conflict intentionality seem to also contain individuals and groups that seek less disruptive alternatives in resolving issues, regardless of their shared negative attitudes toward any given water-related activity.

From a gender standpoint, this trend was most palpable in the female respondents in each town: Whereas the male respondents were visibly concentrated in specific parts of the CCI Scale, the female respondents were more uniformly distributed across the spectrum. In terms of conflict intensification, the data suggests that respondents in Glenorchy were more likely to escalate their conflict engagement, irrespective of gender. In this regard, a third of the respondents (5 female and 6 male) reported being willing to become involved in litigation procedures, reactions that were mostly triggered by water chlorination. This suggests a type of gender-neutral consensus in Glenorchy over the perceived negative impacts of water chlorination, and over the decision-making process behind it. The only other comparable gender-neutral response was seen in Ashburton, but in this case, responses were concentrated at the two lowest levels of the CCI Scale (7 female, 4 male). This suggests that, even though these respondents may have regarded water bottling and/or water chlorination as incompatible, they trusted their water authority infrastructure enough to voice their discontent over those activities through them. The comparative gender distribution in relation to conflict intentionality in Ashburton and Glenorchy is summarized in Table 6.2.

Taking into consideration the various types of conflict intentionalities that emerged in the examined cases, one cannot assume that intense hydropolitical conflict inclinations are widely shared within communities, nor that groups who engage in high-intensity conflict events are fully representative of the views of the whole community, at least not from a *modus operandi* perspective.

Secondly, conflict over the commercialization of freshwater remains a realistic possibility despite low numbers of individuals willing to engage in high-intensity political action. As follows from the data, respondents from Ashburton and Glenorchy

Table 6.2 Gender distribution of hydropolitical conflict intentionality in Ashburton and Glenorchy

Type of Action	Ashburton		Glenorchy	
	Female	Male	Female	Male
Small scale acts of violence, protests, vandalism	4	0	0	0
Litigations, appeals of administrative actions	1	0	5	6
Fines, proposal and permit denials, halting negotiations	0	0	0	0
Petitions, withdrawal of third-party support	2	1	2	0
Delays, report reviews, voicing opposition, editorials	5	3	9	1
None	4	5	3	5

did not report a will to engage in intense hydropolitical conflict in large numbers. Despite an undeniable concentration of high-intensity conflict intentionality over chlorination in Glenorchy and over water bottling in Ashburton, those cohorts do not reflect a plurality of support (11 out of 31 in Glenorchy and five out of 25 in Ashburton). The data, however, does suggest that conflict intensification can still occur despite a minority of locals willing to engage in what can be considered as intense disputes. The majority rule that typically applies in other democratic governance arenas does not fully explain the intensification of water disputes. This is important to consider because decisions over water access and use that appeal to a wide majority of stakeholders and citizens may still bear high risks of hydropolitical conflict intensification. Therefore, it is important for the water authority system to build the irrefutability of its processes in a way that leaves little to no space for residents to engage extra-institutionally in conflicts over its decisions.

The above finding aligns with recent research by Chenoweth, who finds evidence suggesting that as small a portion of the population as 3.5% can trigger the realization of non-violent conflict (2017). Although only five respondents in Ashburton reported a willingness to engage in intense hydropolitical conflict over water bottling, they nevertheless succeeded in materializing the most significant conflict action against any of the water practices examined in this study. Similarly, and even though less than half of the respondents from Glenorchy expressed a will to intensify their conflict engagement, they nevertheless succeeded in materializing litigation actions that fall within the purview of an intense hydropolitical response, with indications that conflict over chlorination could intensify further. In both cases, plurality in conflict intentionality did not adequately reflect the real water conflict risk profile in each community.

Conclusions and Final Thoughts

It follows from the data that water bottling and water chlorination exerted an influence on the emergence of hydropolitical conflict intentionality when respondents espoused concerns over the generation of water issues in their communities. However, two other processes could be discerned, both of which relate to the containment and intensification of hydropolitical conflicts. Whereas the emergence of hydropolitical conflict intentionality was motivated by concerns over water, their containment was facilitated by respondents regarding the decision-making and appeals processes behind each practice as sufficiently robust and legitimate. The risk of intensification, however, increased when respondents regarded these processes as inadequate or illegitimate.

The findings presented in this chapter shed an important light on the central role of state institutions in maintaining effective channels through which hydropolitical conflicts can be ventilated and resolved. They also provide a small insight into the type of hydropolitical conflicts that can emerge in situations where the trustworthiness and transparency of water-related processes and decisions are questioned by communities.

Even though this study examined hydropolitical conflicts in relatively small rural environments, they are not standalone, isolated episodes of discord and animosity. As time progresses, New Zealand continues to witness an undeniable upsurge of intense hydropolitical conflict events which requires the continued attention of both the research and policy communities. If the escalation of hydropolitical conflicts is indeed anchored on communities' trust in their water authorities' decision-making and appeals processes, then one can extrapolate from the findings in this study that water authorities in New Zealand could be facing a progressive decline of trust in their RMA-mandated duties, which could translate into communities preferring extra-institutional arrangements to sway water-related policy-making in their favor.

The above-mentioned decline of trust is already resulting in the intensification of hydropolitical conflicts outside of Ashburton and Glenorchy's contexts. Intense marches over water bottling in Belfast (Canterbury) have succeeded in drawing residents by the thousands (Radio New Zealand, 2019); some of these gatherings have required the intervention of law enforcement officials (Stuff, 2018). Along with their ability to draw strong popular support, protest movements over water-related decisions have succeeded in developing robust financing campaigns to help sustain their conflict actions, including challenges in court. One noteworthy case is the recent statement by the Christchurch City Council, which announced the granting of $50,000 to cover the legal expenses of the Aotearoa Water Action organization (AWA) (Christchurch City Council, 2019). AWA has been challenging Environment Canterbury's approval of two resource consents for water bottling, claiming the decision-making process behind each application breached the RMA. This case illustrates the extent to which local groups are willing to intensify and sustain their actions over untrustworthy decision-making processes. It also sheds some light on the types of incompatibilities inherent between city/district councils and regional councils over water issues, and the extent to which groups coalesce behind water issues.

Like Glenorchy, water chlorination has also motivated mobilization by some groups in New Zealand, highlighting the types of challenges involved in the contemporary implementation of water treatment strategies in the country. For example, Māori groups in Hawke's Bay have protested before city councilors, expressing their discontent with the introduction of water chlorination, and arguing that, as *kaitiaki* (guardians) of water, they opposed the use of chlorine (New Zealand Herald, 2018b). This incident demonstrates the extent to which some Indigenous groups regard water treatment strategies as detrimental to their value systems. Also, in Hawke's Bay, a group referring to itself as Guardians of the Aquifer organized a petition calling for a referendum on the implementation of chlorine in the city of Napier, arguing that chlorine posed undisclosed health risks (New Zealand Herald, 2018a). In addition to these examples, there is evidence suggesting that universal water treatment has had unforeseen negative effects on some industries, increasing the costs of production for enterprises such as breweries and wineries, whose processes are adversely affected by the presence of substances such as chlorine in the water (New Zealand Herald, 2016).

The findings documented in this study, and the examples above demonstrate one crucial truth: the Coleridge Dilemma seems to be a reality in New Zealand, where the country's traditionally abundant freshwater stock does not guarantee that water relations will remain peaceful. Indeed, problems have been identified related to how the existing water authority system can contain conflicts within its local midst.

Beyond the dynamics examined above, this study also uncovered a small segment of participants who did not engage in any type of conflict action within the CCI Scale. Instead, these respondents chose to start enterprises more closely related to community-level cooperation over water. Other respondents also reported identifying the need to create spaces for dialogue over water, in order to influence the betterment of water-based outcomes in their districts. While these participants constitute a small minority of accounts, these stories provide an important contrast to those documented in the previous chapters and merit space within this study. The next chapter will relate these stories and will showcase the opportunities identified by some communities in New Zealand in the face of water-related adversity.

References

Chenoweth, E. (2017). It may only take 3.5% of the population to topple a dictator—with civil resistance. Retrieved June 25, 2019, from https://www.theguardian.com/commentisfree/2017/feb/01/worried-american-democracy-study-activist-techniques

Christchurch City Council. (2019). Council to help Christchurch water group with legal costs. Retrieved April 15, 2019, from https://cccgovtnz.cwp.govt.nz/news-and-events/newsline/show/3530?fbclid=IwAR3-g36rq9L90l3v8LkrIUzYYrUIvyWH9Zp9MaIAtIMTqWtPnjwFzE5CpS4

Government of New Zealand. (2015). Environment Canterbury (Transitional Governance Arrangements) Bill. Wellington.

New Zealand Herald. (2016). Wineries and breweries use filters to combat chlorine. Retrieved April 15, 2019, from https://www.nzherald.co.nz/nz/news/article.cfm?c_id=1&objectid=11745145

New Zealand Herald. (2018a). No chlorine-free water coming soon to Hawke's Bay. Retrieved April 15, 2019, from https://www.nzherald.co.nz/water/news/article.cfm?c_id=362&objectid=12109018

New Zealand Herald. (2018b). Chlorine protest at Hastings council drinking-water pop-up. Retrieved April 11, 2019, from https://www.nzherald.co.nz/hawkes-bay-today/news/article.cfm?c_id=1503462&objectid=12011197&ref=rss

New Zealand Parliament. (2010). Commissioners needed to fix Canterbury water. Retrieved June 1, 2019, from https://www.beehive.govt.nz/release/commissioners-needed-fix-canterbury-water

Radio New Zealand. (2019). Protesters seek halt to water exports. Retrieved April 11, 2019, from https://www.radionz.co.nz/news/national/384343/protesters-seek-halt-to-water-exports

Stuff. (2018). Five arrested after protesters turn off water to Canterbury's regional council HQ. Retrieved April 11, 2019, from https://www.stuff.co.nz/the-press/news/109510593/protesters-turn-off-water-to-canterburys-regional-council-hq

Chapter 7
Conflict Intentionality Classification

Abstract This chapter provides an overview of the types of conflict parties that emerged as a result of the realized or potential introduction of water bottling and water chlorination in Ashburton and Glenorchy. Three types of conflict parties were identified throughout the development of the research and include Revisionist, Status Quo, and Reclusive. This chapter also explains the rise of parties that refuse to engage in any type of water-based conflict and instead choose to initiate different kinds of unilateral, Collaborative actions to help resolve water-related problems in their communities.

Introduction

As a result of the emergence, containment, and intensification processes described in the previous chapter, this study identified four types of parties, which will be explored in this chapter: Revisionist, Status Quo, Reclusive, and Collaborative. These individuals espoused discernable differences in the type of political action in which they would potentially be willing to engage. This chapter will build on the examination above and will provide a taxonomy of the conflict parties that emerged in response to water bottling and water chlorination in Ashburton and Glenorchy.

To do so, this section will provide a categorization of parties based on their conflict intentionality and the intensity of their preferred political action. Additionally, this chapter will provide an analysis of individuals who reported no conflict intentionality of any type and intensity, but who nevertheless expressed a willingness to establish cooperative initiatives to help improve a visibly adverse situation. As was expressed at the beginning of this chapter, one must take note that the analysis in this chapter is based on the respondents' views over their own potential conflict engagement. Unless it is explicitly stated, this analysis does not account for actualized conflict engagement by the respondents.

A. E. Suazo, *The Risk of Water Conflicts in Aotearoa-New Zealand*,
https://doi.org/10.1007/978-3-031-06660-3_7

Revisionist

Firstly, one can highlight a respondent cohort that was likely to engage in intense hydropolitical conflict events, an intent that stems from a profound lack of trust in the water authority regime. The intentionalities expressed at this level reflected a significant skepticism toward the established political paradigms and did not seem likely to lead to any type of engagement or 'reaching out' to elected officials, civil service, or decision-making institutions in general. These respondents seemed the most likely to regard their water authorities as inefficient, unreflective of their needs, and as overly sympathetic to commercial interests, a point that was most salient in Ashburton, but that was not entirely absent in Glenorchy. Additionally, these participants were the most likely to quote Indigenous knowledge and wisdom when framing their perceptions of nature and humans' interactions with elements in the environment. Due to a profound skepticism toward how water is treated and governed, these respondents were the most disposed to disengage from mainstream political fora and the most likely to report being willing to follow disruptive routes to achieve the change they believed they deserved. This type of respondent will be regarded henceforth as Revisionist. Within the context of this study, revisionism relates to actors justifying radical reconfigurations in decision-making and institution building through political actions conceived to disrupt the existing political Status Quo.

The revisionism described above was evident in the account reported by one of the main leadership figures of the Bung the Bore movement, responsible for the 2016 protests in Ashburton over water bottling. When asked about their long-term objectives regarding how water is managed and governed in their district, they argued for the creation of a New Zealand-wide fund that would accrue revenue nationally, but that would be available for the use of small local councils. This fund, they argued, would allow local councils to invest in better water management infrastructure, regardless of the point of origin of said funding. This proposed amalgamation of funding capabilities, and the arrangements to administer said funding, demonstrates the type of actions for which a Revisionist party is willing to advocate. Their views of change are contingent upon significant recalibrations of the existing political order and, in some instances, on the redrafting of jurisdictional authority over water management, governance, and financing. Thus, minor policy changes, and superficial legislative amendments are unlikely to satisfy a Revisionist party's demands for change.

The *modus operandi* chosen by Revisionist actors reflects the notion that their profound views of radical structural change are unlikely to be met by their authorities, which in turn renders them more likely to support alternative, extra-institutional channels to express their discontent. These types of actors, however, are not widely seen as positive bearers of change by other members of the community, and in several cases, Revisionist parties were subject to significant contempt. In Glenorchy for example, one respondent referred to the difficulties experienced by another community member, who is well known for their vocal environmental activism. In their account, they argued that the person in question has had to live with the negative impacts of their political activities, the most significant being their inability to secure stable employment in Glenorchy. Another participant, also from Glenorchy, described a local sustainability-based organization as a 'ginger group', due to its activism to improve environmental standards in Glenorchy, and their recent actions against water chlorination. Similar views were documented in Ashburton regarding the reputational standing of some locals, whom several respondents considered as extremists and as dismissive of facts. Some of these views were so negative that, in some cases, participants who would be considered Revisionist reported having received death threats by members of the community who disliked or disagreed with their activities. In most cases, Revisionist actors seemed cognizant of these consequences and were willing to take responsibility for them. In Ashburton, one respondent who reported being heavily involved in protesting in the district argued that environmental issues are so important that they are 'worth dying for', while a second respondent, also from Ashburton, acknowledged that their political activism was directly responsible for a decline of their health.

Status Quo

In contrast, a respondent cohort reported a willingness to engage in low-intensity hydropolitical action. Even though this cohort espoused similar views, concerns, and objectives as their Revisionist counterparts, they expressed a pursuit for positive change through actions sanctioned by the water authority system. In this regard, the transformative character of their intentionality reflected some type of trust in the system, something that was not as readily manifested within the Revisionist cohort. Furthermore, the positive resolution of disputes in this group remains within the confines of the prescribed political order and its processes and in no way seeks to reconceptualize, reconstitute, or defy the inner mechanics of decision-making. This cohort also seems unlikely to intensify their hydropolitical conflict action in the face of an unsatisfactory decision by policy-makers. These actors will be referred henceforth as Status Quo and will be defined as a party that, notwithstanding its concerns over any one water-related decision, seeks access to and interactions with its water authorities to influence corrective measures.

One good example of a Status Quo account was documented in Glenorchy, where a respondent expressed significant concerns over receiving an order by the Otago Regional Council to install a water metering device in their property. During their interview, the participant shared copies of letters they wrote to the authorities in question, requesting that the order be reversed. Several months after the data collection was completed, the same participant shared additional follow-up letters, where they continued to make the case against water metering, despite their requests being consistently met with unsatisfactory results. This respondent also reported being willing to start similar actions with regard to both water bottling and water chlorination.

Along similar lines, a respondent from Ashburton argued that several residents in their district chose to voice their disapproval of any given water-related issue by removing their support for incumbents during electoral contests, which, in their view, happens often in their district. In this instance, such a removal of support does not question the integrity of the political or electoral systems, and the action focuses solely on the holders of public office. Under no circumstance are these intentionalities envisaged to disrupt the institutional Status Quo, unlike the Revisionist parties described above, whose actions are meant to fall outside of the prescribed political order.

These accounts reflect consistent dissatisfaction with water-related decision-making. They also highlight a preferred *modus operandi,* manifested in the respondents' continued willingness to engage with their decision-making institutions, despite some dissatisfaction. In this regard, Status Quo actors may still espouse profound doubts over the individuals who hold elected office, but these doubts do not translate into institutional and procedural trust being breached. This in turn makes them unlikely to intensify their conflict intentionality, or their potential engagement in intense conflict.

Reclusive

This study also identified a significant number of respondents who reported unwillingness to take political action against water bottling or water chlorination, irrespective of their disagreement with any given water-related issue. Some respondents, for instance, expressed a disinclination to engage in any type of political action, either through their elected officials, or by participating in intense political action. These respondents seemed the most likely to resign themselves in the face of perceived systematic injustice and to seek no action to address the source of said injustice. From this point on, this cohort will be referred to as Reclusive, in acknowledgment of its proclivity to remove itself from any one contested affair.

Reclusive respondents in both Ashburton and Glenorchy seemed to base their reported lack of political responsiveness on the unlikelihood that their potential action could improve an adverse circumstance; the ostensible unattainability of positive outcomes was the main driver of inaction for this cohort. One respondent from Ashburton, for example, argued that they anchored their lack of engagement on the fact that, in their view, their district's water authorities were too friendly with strong actors in the community. Any action to try to reverse policies that favor said actors, they continued, would prove fruitless, as wealth generation in their district is overly dependent on those actors. This type of account was also echoed by a participant in Glenorchy, who expressed that activism was not compatible with their personality, and even if it were, it would not succeed in reversing what they perceived as favoritism toward commercial interests. This participant was referring to the significantly lower water rates in areas such as Queenstown, one of the main hubs of the tourism industry in New Zealand's South Island. They expressed discontent over this asymmetrical policy because, in their view, the sources of Queenstown's freshwater emanate from Glenorchy, and therefore, the locals in their town should benefit from comparatively better water rates.

When asked about their views on water bottling and water chlorination, Reclusive actors often related a significant amount of knowledge with regard to each practice. They also espoused an understanding of the regional, national, and, in some cases, international dimensions behind each enterprise, which seems to have facilitated a perception that the introduction of these practices in their own towns was inevitable. For example, some respondents from Glenorchy referred to the strife over water bottling in the Canterbury region to illustrate the inevitable introduction of water bottling in their district. In those accounts, respondents made references to international demands for bottled water, and to water scarcities abroad, and argued that sooner or later those demands would translate into water bottling becoming a reality in their town, given its significant surplus of high-quality water. Similarly, respondents from Ashburton seemed quick to note that industry-driven water degradation in their district, and elsewhere in New Zealand, was such that policies like water chlorination could not be rolled back, and political action to eliminate those policies would prove useless. The perceived inability to influence an inevitable outcome seemed to inform Reclusive actors' decision to remove themselves from engaging in any type of hydropolitical conflict action.

Cooperation in the Face of Water-Based Incompatibilities

Chapter 6 explained the conditions likely to influence the emergence, containment and intensification of hydropolitical conflicts in Ashburton and Glenorchy. Building on these findings, the present chapter then provided a detailed description of the types of conflict intentionalities identified in both locations. These findings highlight the challenges faced by communities in New Zealand as they cope with the introduction of commercial water practices they regard as deleterious and as they seek alternative means to address these issues. The following section, however, departs from the question of hydropolitical conflicts and endeavors to analyze parties that were unlikely to engage in such episodes.

Throughout the data collection, cases and stories related by locals in both Ashburton and Glenorchy suggested that some community members took it upon themselves to take unilateral action to help improve an adverse water-related situation. In these accounts, locals related a willingness to disengage from institutions, processes, and debates in their communities that they argued as inherently divisive, and instead, decided to establish local networks of Collaborative action. Even though the number of such accounts constitutes a small minority of cases, they provide a beacon of hope, for they are manifestations of what can be accomplished when community members engage with one another in response to the potential for water-based conflicts.

This brief section includes these stories separately, because they provide a stark contrast to the accounts described previously. Parts of this analysis relate directly to the introduction of water bottling and water chlorination in Ashburton and Glenorchy, but it also includes Collaborative episodes motivated by other factors.

Collaborative

The previous chapter provided a detailed overview of the conditions in which water bottling and chlorination conflicts could emerge and intensify in Ashburton and Glenorchy and demonstrated that hydropolitical conflict intentionalities were likely to escalate when local communities espoused a lack of trust in the approval and appeals processes for any given water activity. In contrast, enterprises established under a trustworthy approval and appeals regime seemed to inspire local communities to ventilate any given discontent or disapproval through the local channels and institutions prescribed by the RMA, thus containing hydropolitical conflict dynamics to the lowest levels of the CCI Scale.

Thus far, the focus of this study has been inherently conflict centric. The data, however, indicates that a small number of respondents were likely to forego conflict engagement at any level and seemed to prefer proactive, peace-based channels for the betterment of any given situation instead.

In the previous chapters, these voices were coded within the No Comment or No Action level of the CCI Scale because indeed, these actors' intentionality did not correspond to the types of actions within the conflict side of the CCI Scale. However, as the data collection process continued, these same respondents elaborated their views on political action, thus uncovering some insights that relate more to local community-level organization, than to intra or extra-institutional conflict. These actors also seemed to recognize that water-related adversities could be reversed through such collaborations. This group of participants will be regarded henceforth as Collaborative.

Unlike the Revisionist parties described in the earlier stages of this chapter, Collaborative parties did not seem likely to challenge the integrity of any political institution or process but recognized the system's inability (or unwillingness) to ameliorate an adverse situation. In some cases, Collaborative actors reported having generated institutions and processes that mimicked those already in place in their district, reconceptualized in such a way that enhanced local actors' ability to influence how water was accessed, treated, and used within this proto-institutional framework.

Chapter 5, for example, related the experience of respondents in the small town of Mayfield, near the town of Ashburton, who found a channel for achieving positive change in proactiveness and local agency, while circumventing the political jurisdiction of their water authority (the ADC). In this example, locals accrued sufficient revenue to create their own community water well, on a portion of land that was donated by one of the community members. This collaboration, which occurred outside of the margins of the existing decision-making institutions and processes, was a response to repeated episodes of water inadequacies in Mayfield, which some respondents argued originated from dairy farming and irrigation activities in the vicinity. This example highlights two important dimensions of Collaborative parties: On the one hand, it stresses these actors' unwillingness to resign themselves in the face of an adverse water-related condition; on the other, it showcases their political agency and their ability to reconceptualize institutions and processes as they 'should have been' had they been administered by their water authorities.

A similar account of a Collaborative actor was given by a respondent in Glenorchy, who explained the measures that some of their neighbors undertook during the aftermath of water chlorination. The respondent shared the experience of a resident in the small encampment of Kinloch, whose freshwater derived directly from a clean source in the mountains, and who did not have access to the QLDC-administered water system. In view of the chlorination of water in Glenorchy, the individual made an open call on social media to offer free water from his/her source to anyone in the community who considered themselves to be affected by the policy. As with the community members in Mayfield, some respondents in Glenorchy circumvented the jurisdictional system of their water authorities to create solutions to problems that, in their view, could not have been addressed through interactions with the mainstream political regime.

The examples above demonstrate the willingness of some community members to establish local networks of support and how, in some cases, these endeavors mirror already-existing institutions and processes within the water authority system. The community-conceived collaborations differ, however, from the mainstream ones, in that the former establish institutions and processes that concentrate greater decision-making power in the hands of the stakeholders involved in these collaborations. In this regard, one can argue that Collaborative parties erect parallel proto-institutions and processes to achieve, through their own means, the change they regard as unattainable through the existing authorities.

In addition to the above examples, this study identified other Collaborative parties that sought to create spaces for debate and conversation of heavily polarizing issues, including water. To some extent, the will to create these spaces speaks to a wider reality, perhaps most applicable in Ashburton and its surroundings, where respondents seemed very uneasy about discussing water-related issues, and where several respondents argued not feeling safe enough to discuss these subjects in public. To bridge these conversational and discursive divides, an artist living in the town of Methven (in the Ashburton district) reported that their art gallery had become the de facto place for locals to engage in divisive debates. They argued this was the case due to the gallery being a space where different interpretative approaches were not only welcomed but encouraged. This same individual facilitated the participation of the only representative of the irrigation industry that was willing to share their views on water. The delivery of their account, however, was contingent upon it taking place in the gallery itself, which demonstrates the extent to which its very existence has provided a venue for the proper ventilation of views and perspectives over issues such as water management and commercialization. It also proves the will of Collaborative actors to seek solutions to perceived problems, all at arm's length from their local authorities.

Like the account documented in Methven, there were other Collaborative actors who argued that the use of art had proven beneficial in bringing what one can regard as parties-at-conflict to converse and exchange views. During the data collection, the Ashburton Art Gallery was hosting a temporary exhibit entitled 'The Water Project', a Collaborative project established by 13 New Zealand artists that 'engages with the complex realities of water in the twenty-first century—as bringer of life and ancestral voice, but also as a contested commodity and saleable resource' (The Water Project, 2018). The curator involved in making the exhibit possible in the Ashburton Art Gallery related the difficulties they faced when seeking approval for the exhibit by the ADC, the entity that administers the Gallery. In their account, they stressed the notion that the availability of venues such as 'The Water Project' was crucial for bringing together parties with conflicting interests, who could then take the opportunity to speak with one another, and perhaps gain greater understanding of each other's viewpoints. On this note, they argued:

> The gallery needs to operate in a neutral fashion. There are commentaries in that exhibition that are very critical, but everyone can come here and feel that their voice and opinions matter. (Respondent 3)

In addition to the exhibit, the Gallery established an education program where children could learn about water-related processes such as aquifers replenishment, leaching, and filtration.

While succeeding in creating a safe conversational space that was previously absent, the example of the Ashburton Art Gallery differed from the previous Collaborative accounts, in that the action took place within the existing institutional purview of the ADC. In this regard, the Collaborative space created by the Gallery was not fully detached from the ADC framework, and, to some extent, one can argue that by providing the necessary funding for the exhibit, the ADC was party to the collaboration in question. The installation of the exhibit, however, was conceived by the Gallery's curator in response to a space they argued did not exist within the ADC framework and that was desperately needed in the community.

Conclusions

This chapter presented an analysis of the conflict intentionalities identified in Ashburton and Glenorchy as a result of the introduction of water bottling and water chlorination. It found that intense hydropolitical conflict intentionality was likely to be expressed by actors espousing Revisionist attitudes toward their water authorities, while respondents who reported not being willing to alter the current institutional Status Quo seemed more likely to express an inclination toward low-intensity, intra-institutional conflicts. Lastly, an array of respondents reported no desire to engage in any type of political action against either one of the practices examined in this study.

Additionally, this chapter analyzed accounts where locals reported not being willing to engage in any type of conflict and chose instead to become involved in local cooperative initiatives to help improve a visibly adverse, water-related circumstance. These actors recognized the shortcomings of the existing institutions and processes but, unlike the Revisionist actors examined during the earlier stages of this chapter, Collaborative parties chose not to defy the institutional composition of these entities. Instead, they proactively created their own and, in so doing, provided greater decision-making power to their communities. In addition to creating these proto-institutions and processes, Collaborative actors were cognizant that discursive vacuums existed in their communities and were willing to create safe spaces where these vacuums could be effectively undone. Above all, the existence of these actors reflects the opportunities that could exist in the face of adverse water-related decision-making and the value these actors and communities bring to the attainability of peaceful coexistence over water.

Reference

The Water Project. (2018). *The water project*. Retrieved April 17, 2019, from https://www.thewaterprojectnz.org/

Chapter 8
Research Intersections, Limitations, and Policy Implications

Abstract This chapter explains the theoretical and empirical contributions of this study within the Abundance Theory literature and provides recommendations for future research. This chapter stresses the need to conduct more qualitative research on the causal impacts of commercial practices such as dairy farming and to apply the study's framework in urban environments. It also calls for more research on the causal interplay between water commercialization and conflict in jurisdictions with higher concentrations of Indigenous people.

Theoretical and Empirical Intersections

This study presents evidence that contributes to the ongoing development of Abundance Theory. It does it first, by generating an understanding of how water-based conflict intentionality can emerge and intensify in a water-rich country like New Zealand. As with most other abundant non-renewable resources, conflicts motivated by the commercialization of freshwater can become a reality when individuals and groups in countries such as New Zealand espouse doubts over the approval and appeals processes behind any given operation seeking to access and use freshwater. This demonstrates that even in a situation of surplus, several economic, environmental, political, and social variables can interact to influence individuals' hydropolitical conflict intentionality, and potentially, their conflict engagement.

And second, this study provides a unique insight into the steps immediately preceding the materialization of hydropolitical conflict in a water-abundant setting. By focusing on respondents' conflict intentionality, this study provides an important perspective with regard to how and why individuals shape their initial views and predispositions toward any given water-related enterprise and how those views in turn can translate into individuals becoming increasingly inclined toward conflict engagement. This observation however comes with a crucial limitation: An individual expressing a strong will to engage in hydropolitical conflict may not necessarily translate that intentionality into action. While the data collected through interviews provides a strong indication of the respondents' conflict intentionality, the

A. E. Suazo, *The Risk of Water Conflicts in Aotearoa-New Zealand*,
https://doi.org/10.1007/978-3-031-06660-3_8

documentary sources provide but a small suggestion of their actual conflict engagement. This limitation however opens novel opportunities for research that examines the progression from hydropolitical conflict intentionality to active hydropolitical conflict engagement within water-rich environments. By establishing, to name one example, a long-term ethnographic investigation of an individual's progression from a reported conflict intentionality into a manifested conflict engagement, research can help uncover the causal mechanisms that influence his/her decision-making process, which can then be mobilized and compared across numerous other units.

From a theoretical standpoint, and considering the overall field of environmental security, this study aligns with the Liberal Institutionalist assertion that conflicts over water can be managed through the creation and proper administration of water management institutions and processes. The proposition is supported by those respondents who expressed an inclination toward low-intensity, intra-institutional conflict around water disputes. Under no condition does the Liberal Institutionalist argument negate the existence of conflict; it argues that its escalation can be prevented through institutional development and trust-building. Water bottling and water chlorination motivated a fair number of respondents to become inclined toward conflict, but several of these intentionalities remained within a low-intensity bracket and within the administrative and political purview of the water authority regime. In these cases, respondents who did not express doubts over the approval and appeals process for water bottling and water chlorination were less likely to intensify their conflict intentionality beyond these arrangements, thus contributing to the containment of hydropolitical conflicts.

This is not to say that interactions within the water authority system yielded results that respondents deemed as widely appropriate or remedial. There were indeed accounts of respondents who reported consistent dissatisfaction with their local water authorities, but these events did not seem to disrupt the trust they invested in their water authority infrastructure in any way. Those who reported a will to remove their support by voting against their incumbent officials, for example, demonstrate an implicit trust in the system that is unlikely to translate into institutional challenges, or into hydropolitical conflict intensification.

Low-intensity conflict intentionality in Ashburton and Glenorchy demonstrates the conflict-assuaging potential of liberal institutions such as city/district and regional-level councils in New Zealand and how local communities are drawn to engage with them in the face of a water-related incompatibility. This study, however, cautions the research community from surmising that liberal institutions will effectively contain all conflicts over water. The trust mechanism to which Dinar alludes (2002), that Oye argues is achieved through the creation and solidification of conventions (1986), and that this study finds influences the risk of water-related conflict intensification, should not be synonymous with the existence of liberal institutions and processes. The Liberal Institutionalist argument applies insofar as the authorities that make decisions over water gain and sustain the trust of the communities affected by their decisions. If trust in these institutions and processes erodes, there is a legitimate likelihood that communities will espouse an inclination toward intense conflict engagement, to influence an improvement to a visibly adverse condition. Beyond the respondents' accounts in Ashburton and Glenorchy, high conflict intentionality elsewhere has materialized into protests, although no record suggests that overt violence has emanated as a result of the above conflict escalation. However, this does not mean that violence over water will continue to remain absent in New Zealand. Caution and foresight must therefore dictate the research and policy communities' future work, with a keen focus on reconstituting, reinforcing, and maintaining communities' trust in the state's water authority system.

While it suggests overall support for the Liberal Institutionalist argument, the evidence collected in this study also recommends revisiting some aspects of it. One such dimension relates to circumstances where the water authority regime itself is a party to hydropolitical conflict. As follows from the documentary analysis, the ADC, ECan, QLDC, and the ORC, at different stages, were all parties to conflicts triggered by water bottling and water chlorination, which seems to have created a vacuum in each district's conflict resolution capabilities. In this regard, the data suggests that the state's expected role as a facilitator and administrator of peace can become that of a potential exacerbator of, or party to, hydropolitical conflict. For some respondents, potential conflict intensification seemed to be the only possible route to improve a perceived condition of injustice or inadequacy. For other participants, establishing Collaborative networks within their own communities became the best option to resolve any given water-related issue. In either case, the state did not seem to be a proactive actor in the positive resolution of disputes.

The above observation is part of a lacuna previously identified by Finger in his critique of the literature's failure to problematize 'the institution' within the purview of environmental politics. In this regard, he observes the following:

> Institutions are simply seen as being neutral instruments in the hands of policies, politicians, ideologies, people, organizations, or even technologies. Rarely are institutions considered to be a problem, let alone the problem when it comes to diagnosing or addressing the global environmental crisis (Finger in Park et al., 2008, p. 34)

If indeed the state is expected to manage conflicts over water when they arise, who is then mandated with the task of diffusing conflict when the state is a party to it? Should peacebuilding capabilities be concentrated at the community level, or at the water authority level? What form should these capabilities take when hydropolitical conflicts intensify? And finally: Should peacebuilding capabilities be administered from without the conflict locality altogether? This last question echoes concerns by Gillet et al. in their analysis of local water conflicts in South Australia, where they recommend that conflict resolution capabilities be relocated from the immediate setting of the conflict (local/regional) to a state level (Gillet et al., 2014).

In addition to the question of institutional trust, this study cannot neglect pointing out the role of unusual weather patterns in influencing some of the respondents' concerns over freshwater quality, quantity, and the allocation policies designed by their water authorities. During the data collection in Glenorchy, it was common to find signs around the town suggesting the high risk of forest fires in the area. Further discussions with the respondents uncovered worries over the unusually dry summer of 2018, where locals took notice of the low levels of seasonal precipitation, and how that had impacted the reservoir levels in their town. Respondents also reported having been served with water restriction notices during this period, something that seemed highly unusual for them. Similar accounts were provided by locals in Ashburton, where respondents expressed significant discontent over preferential water allocation in their district during times of drought.

Even though this study's research objectives were not concerned directly with the impacts of climatic change on the likelihood of water-based conflicts in New Zealand, climate-driven uncertainty cannot be denied as a factor that altered, to some degree, the respondents' traditional perceptions over water availability and quality in their towns. In this regard, this study finds support for what several analysts call the 'threat-multiplier' argument (Vivekananda et al., 2014), which suggests that climate change is likely to exacerbate the negative effects of extant state fragilities, bolstering the likelihood of conflict. Even though the argument is normally applied in weak socioeconomic settings, there are some applications of it in Ashburton and Glenorchy, where climate change seemed to have influenced the water access and use capabilities of some of the respondents, leading to pressures of different type and scale. Concerns over the impact of climate change on intrastate water conflicts have already been envisaged by policy-makers. For example, former British Secretary of State for Energy and Climate Change Edward Davey argued that changes in rainfall could impact the robustness of existing water sharing treaties and agreements, increasing pressures between and within states (Government of the UK, 2012). If indeed the erosion of trust in New Zealand's water authority system is partly influenced by changing climatic conditions, then one can conclude that such fluctuations are likely to continue influencing how and why individuals and groups in New Zealand could espouse an intent to engage in hydropolitical conflicts.

The above elaboration highlights important issues related to liberal water authorities' ability to conceive of access, allocation and use policies and processes that account for their conflict-inducing potential. This study uncovered evidence suggesting that water-related institutions and processes in New Zealand, while promoting sound environmental well-being and protection mechanisms, seemed ill-equipped to manage conflict escalation. This finding was perhaps most palpable in Glenorchy, where the otherwise benign decision to securitize the town's freshwater supplies through chlorination induced a relatively unexpected backlash within the respondent cohort, which motivated some participants to report high conflict intentionality and, in some cases, active conflict engagement. In this regard, attention must be provided to develop and bolster conflict resolution capabilities at the district and regional levels, as they are the governance structures mandated with implementing socially, economically, and environmental-sensitive water management strategies.

A reasonable way to bolster these capabilities is by conflict-proofing the parts of the water authority system that are in direct control of local authorities, such as the resource consenting process that councils are required to follow to approve the introduction of commercial operations. In this regard, the consenting process could account for water-based conflict risk at two levels. First, when making decisions over a resource consent application, local authorities could request applicants to prepare a form of conflict risk assessment to determine the extent to which an activity could motivate disputes within communities. This assessment would be broader than the current practice of identifying potential Affected Parties and would be more focused on the community-wide economic, social, and environmental compatibilities of a proposed activity. This assessment could strengthen the evidence base that local authorities use to make consenting decisions under the RMA and would help to proactively identify potential sources of disputes.

And second, should an application be successful, councils could require consent holders to report on the emergence of water-based conflicts as a consent condition. This could include reporting on early signs of conflict, in line with the least-most intense conflict events coded in the CCI Scale. It is expected that such reporting requirements could help improve local authorities' ability to identify the emergence of disputes and enhance their ability to intervene should it be required. This in turn is expected to have positive impacts on the prevention and management of conflicts related to approved resource consents and on the enhancement of trust in the system.

Beyond the theoretical application of the Liberal Institutionalist framework, this study documented cases where water-driven conflict intentionality (and some manifested conflict cases) in Ashburton and Glenorchy intensified when commercial operations were approved under dubious and untrustworthy decision-making processes. In this regard, the causal mechanism applied by Neo-Malthusianism theorists—competition—seems to find some application in the case studies examined in this study. This application is worth noting because the competition mechanism that Neo-Malthusians argue emerges from natural resource scarcity (Homer-Dixon, 1999) and also seems to manifest in resource-rich environments like New Zealand's. In this regard, Abundance Theory seems to share some causal patterns with the Neo-Malthusianism framework.

The competition dimension was most evident in the town of Ashburton. For example, there is evidence suggesting that the introduction of water bottling in Ashburton would have increased the number of actors competing for water access and use, with a significant number of respondents perceiving themselves as placed in a comparatively disadvantageous position by a flawed decision-making process. The reported participation of members of the irrigation and farming communities in the 2016 protests further suggests that industry-level competition was also palpable in Ashburton, and that several economic actors saw the introduction of a water bottling operation as an activity likely to increase water claims in the district. This study suggests that the competition mechanism presented by Neo-Malthusianism theorists is not the exclusive manifestation of a resource scarce environment: Competitive behaviors in a water-abundant setting can still be triggered when local communities and interest groups regard a water commercialization practice as a source of problems. The Neo-Malthusianism framework, however, does not apply any further beyond the above manifestations of competitions.

From an empirical standpoint, this study complements a series of different research findings in the environmental security literature. Perhaps the most striking empirical alignment occurs with Doppelhoffer et al. and their examination of mineral resource abundance and conflict. As per Chap. 1, Doppelhoffer et al. argue that the abundance of minerals is correlated with political instability and with what they term rent-seeking behavior (2002). Reports of district-level disruptions motivated by the introduction of water bottling and water chlorination evoke some of the concerns raised by Doppelhoffer et al., although the instability to which they allude manifested differently in Ashburton and Glenorchy, where it remained confined to non-violent forms and expressions of hydropolitical conflict. In addition to this, and beyond the data collected for this study, water abundance seemed to motivate a steadfast growth in the number of actors seeking to draw an economic benefit from their access to and use of freshwater, evident in the growth of industries such as dairy farming, irrigation, tourism, and water bottling in Ashburton, Glenorchy, and arguably in the rest of the country. Therefore, the political instability and rent-seeking behaviors that Doppelhoffer et al. argue emerge from the abundance of minerals and also apply as manifestations of freshwater abundance.

This study moreover contributes to work that examines the causal relationship between natural resource abundance and democratic performance. For example, Jensen and Wantchekon argue that natural resource abundance contributes to weaker democratic performance, which can then translate into the emergence of conflict (Jensen & Wantchekon, 2004). They argue this effect is caused by the rise of vertical decision-making institutions seizing too much discretionary power over abundant natural resources (2004). Indeed, this book identified a Revisionist respondent cohort whose distrust toward water authorities motivated its inclination toward high-intensity, extra-institutional conflict intentionality. Beyond these participants, distrust in the local elements of the water authority system may also be a contributing factor in New Zealand's generally declining participation in local government elections (Local Government New Zealand, 2019). Together, these findings suggest that local communities may be becoming progressively reluctant to seek access to their liberal institutions and, to some degree, are becoming less inclined to interact with, or contribute to, the democratic life in their districts and regions. Poor citizen participation and input, one can argue, only contribute to the further deterioration of the elected water authority system in New Zealand; this aligns with the theoretical proposition presented by Jensen and Wantchekon that natural resource surplus leads to negative outcomes, in this case, manifested in poor democratic performance.

This study also documented, albeit from the periphery, a small number of cases where respondents engaged in informal arrangement to develop solutions to water-related problems. Analysis conducted on the amount of unregistered water supplies in New Zealand suggests that such informal arrangements indeed abound across the country, with an estimated 75,000 of them being identified (Taumata Arowai, 2022). To some extent, these informal institutions and processes constitute a reimagined version of their water authority regime, at least in part, where decision-making power and stakeholdership are devolved to the water users themselves, and where safe havens for conversations over water are created outside of the water authority's purview.

The installation of art galleries, community water sharing, and Collaborative water infrastructure investments are clear examples of the type of cooperative frameworks that some respondents argued were best positioned to improve an adverse water-related condition triggered by the introduction of water bottling, water chlorination, or the continued development of industries such as dairy farming and irrigation. However, these findings constitute peripheral observations, which require deeper examination to accurately assess the mechanisms and processes that allow for such cooperative behaviors to materialize.

To garner a better understanding of how cooperation can be fostered, peace, and not conflict, should be the guiding principle driving future research questions. This is a concern previously raised by Shrivastava and Suazo in their exploration of what they call 'Peaceful Sustainability'. In their examination, they say the following:

> Peace-centered sustainability endeavors require reformulating research questions into enquiries, which are inherently peace-focused: How does environmental degradation affect peace and sustainable development? How does environmental health affect positive peace infrastructures? Answering such questions will inevitably lead our scientific community to identify stakeholders for peace in both sustainable development and human security circles, and thus arrive at dynamic solutions that effectively incorporate human and environmental dimensions. (2017, p. 31)

Examining such questions would inevitably lead to the study of what Amster calls 'Peace Ecology', which he defines as follows:

> [An environment] in which people possess tools for resolving conflicts and restoring relationships, distributing resources and opportunities in just ways, and promoting the health and wellbeing of all constituents. It is also a society that relates to the balance of the biosphere in positive and healthy ways, that limits its ecological footprint and sees itself as part of nature rather than its superior. (2015, p. 20)

Limitations

The above theoretical and empirical exposition demonstrates the complexity inherent in the management of abundant freshwater resources in New Zealand and the extent to which different types of commercial operations are likely to inspire communities to become inclined to start and intensify conflicts. It also proves the challenges the research and policy communities face in properly assessing predictable patterns of behavior over water access and use in New Zealand. Environmental, economic, social, and political heterogeneity seem to influence the types of allowances (or lack thereof) that New Zealand communities are likely to provide in relation to water-based conflicts caused by a commercial water practice.

In this regard, this study devoted its resources to the analysis of two rural environments in New Zealand which, to some extent, highlights the challenges inherent in the introduction of water bottling and water chlorination in areas with similar conditions to those found in Ashburton and Glenorchy. There are, however, indications that water conflicts continue to emerge and intensify in urban areas such as Christchurch; this suggests that the water incompatibility and institutional trust mechanisms may also be applicable in those environments. To accept this assertion as a given, however, would be premature, as rural and urban centers diverge significantly with regard to their water access and use criteria. Therefore, future research needs to dedicate adequate time and effort to examining the commercialization-conflict dyad as it applies in urban centers.

In addition to expanding this study's framework beyond its rural context, further research needs to assess how the commercialization of freshwater may influence the emergence and intensification of hydropolitical conflict in Māori communities. Ashburton and Glenorchy do not have significant Māori populations, which renders the results of this study difficult to replicate in more densely populated Indigenous areas. Additionally, this study did not capture a sufficiently representative sample of Māori respondents to adequately reflect Indigenous views and interests over freshwater and hydropolitical conflict. With this in mind and considering the special status of Māori in New Zealand, this study urges the research community to devote special attention to how the commercialization of freshwater may create conflicts akin to those found in Ashburton and Glenorchy in communities with greater concentrations of Māori residents. Such efforts will inevitably require special respondent identification strategies to adequately target Māori participants in the location of interest. They will also need to develop strong networking capabilities through known Indigenous leaders (including academics), groups, organizations, and policy-makers to maximize their ability to secure adequate Māori participation.

While examining the causal effects of water bottling and water chlorination proved beneficial from a theoretical and empirical standpoint, this study does not directly allude to the specific impacts of the dairy industry, a significant source of water-related incompatibilities in New Zealand. Indeed, some accounts presented in this book suggest that the intensification of the dairy industry has influenced certain communities to remove their trust in their water authority regimes, which, by application of the arguments presented in previous chapters, would suggest that these communities are likely to engage in intense hydropolitical conflicts over dairy farming. Cases of protests, such as the one presented at the beginning of preface over dairy intensification in the Mackenzie River, prove the wider application of this study's arguments beyond the initiation of water bottling and water chlorination projects. This does not, however, delve deep enough into the specifics of the dairy industry, the conflict parties involved, and the incentive structures that have allowed for the industry's rapid growth. Therefore, this research recommends a proper examination to determine whether the arguments that apply to water bottling and water chlorination are indeed translatable to the dairy industry.

Lastly, and bearing in mind the conflict-inducing potential of water chlorination, this study recommends that further research be done on the connections between water securitization strategies and the emergence of hydropolitical conflict. This recommendation emanates from the findings in the town of Glenorchy, where water chlorination motivated some of its residents to develop inclinations toward hydropolitical conflicts of different intensities. This study raised doubts over New Zealand's country-wide embrace of water securitization strategies and treatment and presented evidence suggesting that not all communities in the country fully accept the merits of solutions such as chlorination, nor accept the decision-making process behind them. As was the case in Glenorchy, chlorination was regarded as a policy that respondents perceived as unnecessary considering the town's traditionally pristine freshwater supplies. By reducing the locals' ability to influence or challenge the chlorination decision prior to its implementation, the QLDC inadvertently motivated some respondents to intensify their conflict intentionality beyond engagements with the submissions regime, prompting some to organize collectively to challenge the decision in the courts and to potentially organize other episodes of higher intensity, including marches and protests. Whether conflict intensification was something the QLDC anticipated remains open for scrutiny. Nevertheless, one issue remains clear: Some communities are likely to regard the implementation of water treatment strategies as sources of incompatibilities. As these strategies are increasingly regarded as urgent, and as pressures to implement them rapidly drive water authorities to forego adequate community consultations, the water incompatibilities outlined above are likely to be seen as a continuous imposition from above, thus motivating respondents to prolong their resistance.

References

Amster, R. (2015). *Peace ecology*. Routledge.

Dinar, S. (2002). Water, security, conflict, and cooperation. *SAIS Review, 22*(2), 229–253.

Doppelhofer, G., Miller, R. I., & Sala-I-Martin, X. (2000). *Determinants of long-term growth: A Bayesian averaging of classical estimates (Bace) approach* (No. 266). *OECD economics department working papers*. Paris.

Finger, M. (2008). Which governance for sustainable development? An organizational and institutional perspective. In J. Park, K. Conca, & M. Finger (Eds.), *The crisis of global environmental governance: towards a new political economy of sustainability* (p. 34). Routledge.

Gillet, V., Mckay, J., & Keremane, G. (2014). Moving from local to State water governance to resolve a local conflict between irrigated agriculture and commercial forestry in South Australia. *Journal of Hydrology, 519*, 2456–2467.

Government of the UK. (2012). *Edward davey speech to climate & resource security dialogue for the 21st century conference*. Retrieved June 26, 2019, from https://www.gov.uk/government/speeches/edward-davey-speech-to-climate-resource-security-dialogue-for-the-21st-century-conference

Homer-Dixon, T. (1999). *Environment, scarcity and violence*. Princeton University Press.

Jensen, N., & Wantchekon, L. (2004). Resource wealth and political regimes in Africa. *Comparative Political Studies, 37*(7), 816–841.

Local Government New Zealand. (2019). *Local elections 2019: Quick facts*. Retrieved February 1, 2022, from https://www.lgnz.co.nz/assets/Elections-Fact-sheet-19.pdf

Oye, K. A. (1986). Explaining cooperation under Anarchy: Hypotheses and strategies. In K. A. Oye (Ed.), *Cooperation under Anarchy* (pp. 1–24). Princeton University Press.

Shrivastava, P., & Suazo, A. E. (2017). Peaceful sustainability. *Business, Peace and Sustainable Development, 10*, 28–34.

Taumata Arowai. (2022). *Frequently asked questions for: unregistered, rural and small drinking water suppliers*. Retrieved February 13, 2022, from https://www.taumataarowai.govt.nz/assets/Uploads/FAQs/FAQs-for-unregistered-small-drinking-water-suppliers.pdf

Vivekananda, J., Schilling, J., & Smith, D. (2014). Climate resilience in fragile and conflict-affected societies: Concepts and approaches. *Development in Practice, 24*(4), 487–501.

Chapter 9
Mitigations Through Reform

Abstract This chapter provides an overview of recent reforms aimed at improving how decisions over water access and use in New Zealand are made by local authorities. These reforms are also likely to impact on the delivery of water-related services, and on their regulations. This chapter explains how these reforms are likely to address the intensification of water-based disputes in New Zealand.

Introduction

Throughout the presentation of this book, it was found that the Coleridge Dilemma indeed has bearing in a New Zealand context: There are conditions that enable the emergence of hydropolitical conflicts, and there is a risk that these may intensify because of the decision-making structures that support the approval of commercial operations. As such, the element of trust has been highlighted as an issue the water authority system in New Zealand needs to reinforce to ensure a full containment of conflict dynamics, as explained in Chap. 6.

The question of trust creates a natural segue into the role of local government as the administrator of the water management regime in New Zealand. As elaborated in Chap. 2, carrying out duties under the RMA is the responsibility of local authorities, which, from a conflict perspective, begs questions in relation to decision-making that is deemed unreflective of local interests.

Throughout the development of this book, water and environmental issues in New Zealand have become increasingly prominent in the mind of the public, and pressures for greater certainty and confidence in the water authority system have motivated a range of profound policy changes and reforms. Together, these changes are expected to initiate generation-long shifts in the environmental management space.

In the context of reforms that impact on water management, the 2016 Havelock North contamination incident described in Chap. 2 was a catalyst of sorts; it demonstrated the extent to which suboptimal infrastructure investment and water quality regulation could trigger large-scale contamination outbreaks. It also raised the profile of water issues to such a level of prominence, that the political class had no other

A. E. Suazo, *The Risk of Water Conflicts in Aotearoa-New Zealand*,
https://doi.org/10.1007/978-3-031-06660-3_9

choice but to initiate significant policy changes to meet the public's increasingly higher expectations.

This chapter introduces these changes and puts them in the context of this study's findings. Even though these changes are primarily rooted in improving environmental regulations, and their drivers are unrelated (at least directly) to water disputes, there are a series of parallel, implicit benefits that intersect with the observations presented in this book. It is noted that several of these policy changes are either planned, or in the middle of their implementation, and it is difficult to ascertain their real impact in the hydropolitical conflict arena. At this stage, instead, focus will be given to how these changes are expected to address hydropolitical conflicts based on their policy intent.

Three Waters Reforms

Following the Havelock North outbreak, the government initiated a full review of New Zealand's Three Waters management framework (Department of Internal Affairs, 2021) and agreed to conduct major system reforms. These reforms are being primarily advanced through the introduction of three pieces of legislation. First, the government enacted the Taumata Arowai—the Water Services Regulator Act in 2020, which led to the establishment of an independent water services regulator with a remit to ensure that all communities have access to safe drinking water (Taumata Arowai, 2022). The regulator is also endowed with an oversight function to protect the environment from the impacts of wastewater and stormwater (Taumata Arowai, 2022).

Second, the government enacted the Water Services Act in 2021, which empowers the new water services regulator with a framework that will enable the performance of its formal functions (Parliamentary Counsel Office, 2022). Whereas the previously mentioned legislation was anchored on establishing the new regulator, the Water Services Act 2021 includes the powers that will support its mandate. These include empowering the regulator to administer the DWSNZ, a function previously managed by the Ministry of Health (Ministry of Health, 2021), and provisions to require currently unregistered water supplies to register with the regulator and follow its standards and guidelines. These supplies would include those covered in previous parts of this study, which were initiated by community members through the formation of informal, alternative water provision models. The Water Services Act 2021 also entitles the regulator to establish a framework for providing transparency about the performance of drinking water, wastewater, and stormwater networks, and their operators (Parliamentary Counsel Office, 2022).

And third, the government is in the process of introducing the Water Services Entities Bill in Parliament (Department of Internal Affairs, 2022). The objective of the Bill is to transfer water service delivery functions (currently held by local authorities), into the hands of four new multiregional entities. It is intended that these entities will have responsibility over water service delivery across all areas in their jurisdiction and will remain collectively owned by local authorities. This change is motivated by concerns that current water service delivery arrangements would be incapable of supporting new regulatory requirements due to insufficient revenue-generating capabilities by several councils. As such the Bill intends to apply the principle of amalgamation so communities can benefit from economies of scale, which will enable improved funding and financing capabilities. It is noted that the value and merits of service delivery amalgamation have inspired similar reforms to the water sector elsewhere, including Scotland (Water Industry Commission for Scotland, 2022).

Even though the Three Waters reforms are aimed predominantly at improving the state of New Zealand's water service delivery capabilities, it inadvertently provides crucial opportunities to address the issues highlighted in previous chapters.

First, from a conflict perspective, reforms to the Three Waters system afford an opportunity for enhancing New Zealand's water-based conflict resolution capabilities. Chapter 8 explained the complexity inherent in resolving water-based disputes that included elements of the water authority system as parties to a conflict. This was indeed the case in the events motivated by water chlorination and water bottling in Glenorchy and Ashburton, respectively. Furthermore, it was also noted that a perceived erosion of trust may lead to a greater number of water-based disputes that include local authorities as parties to said conflicts. Through the Three Waters reforms, establishing an independent water services regulator means that a neutral party now exists that can serve a conflict resolution function when required. This constitutes an implicit displacement of dispute resolution capabilities from a local setting to a centralized, more capable one. This displacement aligns with research by Gillet et al. who highlight the merits of such centralizations for improving conflict resolution capabilities (2014). Added to this, the regulator's independent character may also afford new opportunities to serve as arbitrator when parties to a water dispute cannot arrive at a natural resolution.

This last point is particularly germane considering the potential introduction of the Water Services Entities Bill. Based on the policy intent that has been publicly shared to date, what the Bill would be achieving from a conflict perspective is reducing the number of potential conflict parties from 67 (the current number of individual district councils), to just four (the intended number of multiregional water service providers). This means that if disputes emerge over water service delivery in the new system, these can be more easily identified and mitigated. Added to this, if the water services regulator is ever called upon to serve de facto conflict resolution functions, the number of stakeholders with which it will need to engage will become drastically reduced. Together, these are likely to contribute to improved water dispute resolution capabilities on the ground.

Second, the Three Waters reforms are conceived with improved transparency and accountability as the foundations of change. If the new system can deliver on this aspiration, communities are likely to be disinclined from engaging in high-intensity conflict based on a renewed trust in a more capable, transparent, and accountable system. This, however, will require thorough examination once the new Three Waters system is fully operational, and when enough time has passed for assessing the new system's conflict resolution contributions.

Third, it is noted that reforms to the Three Waters system are likely to align with the visions of change presented by both Revisionist and Status Quo conflict parties. As described in Chap. 7, a Revisionist conflict party is likely to engage in intense hydropolitical conflict action based on their profound distrust of the water authority system. Their views of change are anchored on profound rearticulations of the system, in line with the proposals in the Three Waters reforms. Throughout the data collection stages of this research, there is evidence that some Revisionist parties would align, for example, with the intent behind amalgamating New Zealand's water service delivery framework: a process that is deeply complex, likely to reconfigure the current state of local governance in New Zealand, and that will fundamentally change how water users interact with their water providers. In this regard, the policy intent behind the Three Waters reforms is as ambitious and as profound as the drivers motivating Revisionist parties to engage in intense hydropolitical disputes. It is expected that this alignment will help decrease the process of hydropolitical conflict intensification elaborated in Chap. 6.

Similar observations could be drawn about the extent to which the Three Waters reforms are likely to contribute to Status Quo parties' continued intra-institutional engagement should they choose to become parties to a dispute. This is premised on the reforms being engineered to enhance water users' trust in the system, and on those who administer it. Through the reforms in general, and through the potential introduction of the Water Services Entities Bill in particular, it is expected that the new Three Waters system will be significantly more responsive to community expectations, be more transparent and accountable, be better resourced to engage with communities and reflect their views and interests in decision-making, all the while remaining collectively owned by local authorities.

Lastly, this study captured views from community members that reflected some of the drivers behind informal water service delivery arrangements. As it was noted in this chapter, these arrangements abound across the country, and to some extent, seem responsive to a consistent failure by some local authorities to provide safe access to water services. Through the Water Services Act 2021, these kinds of water provision schemes will need to become registered under the new water services regulator and follow its standards and guidelines. Even though the high proportion of currently unregistered water supplies poses logistical challenges in terms of registration procedures, timeframes, and reduced capacity and capability by the smallest supplies, their registration and formalization provide important opportunities for the new system to build improved trust. Alongside these changes, and pending the successful introduction and implementation of the proposed Water Services Entities Bill, it is likely that the Three Waters Reforms will produce a system that is better capable of satisfying consumer demand, and providing consistency in service delivery quality. This is likely to translate into community members becoming disincentivized from seeking alternative water provision options.

Some Challenges Ahead

Despite the benign policy intent driving the Three Waters reforms, their general scope, and particularly that of the proposed Water Services Entities Bill, has been recently faced with challenges by several parties, including a group of 31 local councils called Communities 4 Local Democracy (Communities 4 Local Democracy, 2021). In their view, transferring service delivery powers and water asset management functions away from local government are ill-suited for providing responsive, accountable water services for communities. Appropriate compensation for asset transfers is also an issue that remains unresolved.

Even though the examination of local government's reactions to the proposed Three Waters reforms is out of the scope of this study, it is worth noting that some of these reactions, and their subsequent mobilizations, can be at least partly explained by some elements in the peace and conflict studies field. For example, through the application of spoiler management theory, groups such as Communities 4 Local Democracy constitute an example of estranged leaders and factions that seek to disrupt a process they regard as incompatible with their own interests, groups otherwise known in the literature as spoilers (Stedman, 1997).

Spoiler management theory also contemplates the existence of outside spoilers: parties who seek to undermine a process because of their exclusion from it (Stedman, 1997). In the context of the Three Waters reforms, the removal of local authorities' water service delivery functions could be interpreted by the latter as an exclusion of sorts where councils, in their view, are becoming removed from New Zealand's water authority system. Because of this perceived removal, it could be argued that some councils see no other choice but to organize strategically to disrupt the Three Waters reforms from the periphery of the process. This aspect of the reforms is encouraged to be further analyzed through a spoiler management lens; this will ensure that the intent behind the reforms is not undermined by perceptions of exclusion within key stakeholders.

Resource Management System Reform

In recent years, pressures have progressively emerged to revise the RMA (New Zealand Centre for Political Research, 2020; New Zealand Herald, 2018; Stuff, 2019) because of views it has not adequately protected the environment, enabled much-needed infrastructure investment, responded to population growth, and future-proofed New Zealand against the challenges posed by climatic uncertainty.

After receiving recommendations from an independent panel, the government has initiated reforms to its resource management system (Ministry for the Environment, 2022). These reforms involve fully repealing the RMA in its current form and replacing it with three separate pieces of legislation: The Natural and Built Environments Act (NBA), the Strategic Planning Act (SPA), and the Climate Adaptation Act (CAA) (Ministry for the Environment, 2022).

The proposed NBA is designed to be the main replacement of the RMA. The aim of the NBA is to holistically protect and enhance the environment in a way that supports communities' current well-being and needs without jeopardizing those of future generations (Ministry for the Environment, 2021). To achieve its intent, the NBA proposes that natural resource access and use comply with environmental limits, to be included in a mechanism called the National Planning Framework (Ministry for the Environment, 2021). The NBA would also consolidate all planning documents in each region into one Natural and Built Environments Plan that would provide the framework for the integrated management of the environment in each region (Ministry for the Environment, 2021).

In parallel to the NBA, the SPA will require the development of long-term regional spatial strategies, to better coordinate and integrate decisions of relevance to the environment (Ministry for the Environment, 2022). And the CAA proposes to address complex issues related to managed retreat, and to building resilience in light of the impacts of climatic uncertainty (Ministry for the Environment, 2022).

Collectively, these new statutes are conceived to modernize New Zealand's resource management system, so it is more responsive to the needs of communities, and to improve how local authorities make decisions over natural resource access and use.

From a water conflict perspective, this reform program is expected to address some of the issues identified in this study. For example, if the final statutes fully reflect the reform program's policy aspirations, the new system is likely to enhance local authorities' ability to make uniform, replicable decisions on water access and use in a way that leaves little space for skepticism in communities. A new resource management system that enables consistently accountable and transparent decisions, and that affords practicable enforcement mechanisms, is likely to contribute to greater trust both in the communities decided upon, and within the decision-making parties in the system. It is expected that these changes will help reduce the risk of hydropolitical conflict intensification over decisions made under the new system.

Reforms to the resource management system also afford unique opportunities to improve the consultation processes that support decisions on natural resource access and use, including water. Indeed, inadequate consultation was an issue that participants raised as contributing to their discontent toward how water-related decisions in their districts were made. Consultation requirements are core components of how resource consents are approved under the current RMA, yet the evidence collected in this study suggests there is space for improvement regarding how local authorities consult with their communities to support decisions under the RMA.

In some participants' views, the current consultation criteria under the RMA are interpreted inconsistently by local authorities. Whereas from a process standpoint, some consultations may satisfy the RMA's requirements, they may not be carried out in a way that residents become convinced their views are appropriately captured and considered. This was certainly a factor that contributed to the intensification of conflict dynamics in the introduction of water chlorination in Glenorchy and water bottling in Ashburton. In contrast, when consultation criteria under the RMA are satisfied through consultation processes that adequately reflect communities' views and interests, the risk of hydropolitical conflict intensification remained low and within councils' immediate control. In these cases, should residents remain in disagreement over decisions informed by these consultation processes, it is expected that the ventilation of their discontent will occur within intra-institutional settings sanctioned by the resource management system.

Considering these reforms, decision-makers would benefit from paying close attention to how consultation requirements in the new system can be strengthened, so it can enable wholesome, consistent interactions between local authorities and the communities they represent. Added to this, significant opportunities exist for the creation of consultative mechanisms that reinforce partnerships and relationship-building with Māori.

The policy intent behind the Resource Management reform is also consistent with the aspirations exhibited by some of the identified conflict parties; it is argued that greater accountability and improved engagement capabilities in a new resource management system could lead to alignments with the expectations of Revisionist conflict parties, as described in Chap. 7. This conflict party cohort was the most likely to reflect dissatisfaction with the current resource consent regime, noting that the current RMA enables, for example, the approval of water permits that are often too long in duration. Water permits approved for durations as long as 35 years pushed some of the participants to become more obstructive in their conflict preference, an issue that was most salient in Ashburton over water bottling.

Alongside shortened water permit duration, reforms to the Resource Management system could contribute to a peace-enhancing environment if it addresses the reversibility of decisions under the new system. Throughout the data collection that supported this book, evidence was gathered suggesting that long water permit lengths, combined with a view that decisions could not be reversed under the current RMA, contributed to the escalation of water conflict dynamics. Addressing the reversibility of decisions under the new system is likely to create an environment where, for instance, Revisionist conflict parties are likely to feel disincentivized from engaging in high-intensity conflict events.

Throughout the data collection process, participants shared some insights into how unusual climate events and patterns impacted on water availability and quality in their districts. The climate-related issues reported in this study could be addressed, albeit on a long-term basis, through the reform program's focus on climate change and resilience. The CAA, for example, is designed to support New Zealand's response to the effects of climate change, which is anticipated to help address the water availability pressures attributed to abnormal climate conditions identified in previous chapters.

Chapter 8 explained the opportunity of conflict-proofing the consenting process under the current RMA, so it could better enable local authorities to manage the risk of water-based conflict intensification. It was argued that this could include requiring conflict risk assessments as part of the consent application process and mandating consent holders to report any perceivable early signs of conflict motivated by the consent in question. With the repealing of the current RMA, there are opportunities to incorporate this conflict-proofing in the very architecture of the new resource management system. For example, the consolidated Natural and Built Environments Plans proposed under the NBA could include provisions that reflect how the risk of water (and resource) dispute intensification can be managed by local authorities in each region. Reform also provides an opportunity to stipulate the levels of governance that can deliver conflict resolution capabilities, and the rules for deploying these capabilities. It is noted that conflict-proofing the new resource management system requires dedicated space and analysis; this study encourages detailed elaborations into how to enable a more conflict-sensitive resource management system.

Lastly, there are opportunities for a new resource management system to improve how Reclusive conflict parties engage with it. For example, and in line with the observation in the previous paragraph, the inability to reverse decisions was one of the main criteria that drove Reclusive parties' poor engagement with local authorities. If the new resource management system is built in a way that enhances decisions' reversibility through consenting requirements that encourage minimal adverse impacts on the environment and communities, and that require more stringent compliance parameters for the keeping of current consents, then the new system would provide few reasons for Reclusive parties to abstain from getting involved in affairs of their interest. This is expected to contribute to local democratic life and foster an environment in which a wider range of participants can become convinced that their inputs are impactful.

New Zealand in a Liminal Space

Change is challenging; it inspires feelings of discomfort from the uncertainty and ambiguity that comes from becoming something new and untested. The promise of change comes hand in hand with skepticism from those whose perceptions of livelihood and dogma are not reflected in the change in the making, at least not explicitly. The recalibrations facing the water sector in New Zealand are substantial, and communities across the country are confronted by the change these rearticulations will entail. This in no way means that the changes sought through the above-mentioned reforms bear ill intent or are meritless: It means that the country is in the middle of facing the discomfort inherent in change.

In a way, New Zealand's water sector can be argued to be in a Liminal Space of sorts: a change in condition that 'involves a temporary suspension in which [a] subject is neither one state nor the other—they are in a paradoxical condition where potentiality is at a maximum and actuality is at a minimum' (Soderlund & Borg, 2018).

Anytime a person, and communities by extension, undergo a process of change 'they go through cognitive reasoning, risk evaluation and emotional stages to either inhibit or promote change' (Jeremiah et al., 2020). The tension between the promise of reform and having to confront the changes it entails is perhaps best described by Jeremiah et al. in their exploration of liminality:

> [There is an] initial energy, excitement, confidence, self-belief and support that many people experience during transitional life-periods. Yet, when unforeseen mental and physical challenges arise, and the tension from states of disequilibrium grows, this very burden of change can shock and unsteady even the most confident of individuals. This tension is derived from experiencing the uncomfortableness of the unknown, feeling displaced, uncertain and not knowing how to navigate out. (Jeremiah et al., 2020).

The discomfort of change is expected to be directly proportional to its scale. Indeed, reforms to the Three Waters and Resource Management systems are no patchwork; they constitute a major departure from the Status Quo, where the problems being addressed in some cases, date back several decades. As a result, the habits, ways of doing and living, and in some cases, the benefits that have been enabled by the Status Quo are very likely to be confronted by the changes proposed by the reform agenda. Simply put, reforming these systems is not meant to be a neat and frictionless process.

However, awareness of where the reform agenda sits relative to the process of change affords useful insights into how best to navigate through the uncertainty inherent in the proposed shifts. It also provides valuable opportunities to preemptively identify sources of skepticism and deploy engagement capabilities in a way that socializes the values of reform with those groups. In this context, entrepreneurial science helps explain the cognitive processes involved in change. For example, Jeremiah et al. argue that any change involves three micro-processes that dictate its realization: separation, transition, and assimilation/integration (2020).

As such, it can be argued that even though Three Waters and Resource Management System reforms are at different stages in the process of change, they are both still navigating at various levels with the initial process of separation from the Status Quo. In other words, reform is still in its most inceptual state. For example, even though some parts of the reforms to the Three Waters system could be argued to be in a form of transitional state, such as the institutional maturing of the new water services regulator, some separations are likely to continue as envisaged by the proposed introduction of the Water Services Entities Bill. In contrast, Resource Management System reform is still in the process of designing the format, scope, and procedure for breaking away from the current system.

As described earlier, the potential for a new Three Waters and Resource Management system to address the emergence and intensification of water disputes is significant. However, these new capabilities are anchored on the reform programs' full realization. Given that reform is still in its very initial stages, it is difficult to ascertain the extent to which its progress may continue, and what roadblocks may be placed in its path in the coming years. This also means there are opportunities to socialize the value of reform and gain a much wider consensus on both the problematic nature of the Status Quo and the benefits of change.

Conclusions

This chapter explained the intent behind the planned reforms to the Three Waters and Resource management systems. It further elaborated on a series of identified opportunities the reforms are likely to deliver collectively, and the extent to which they will help address some of the issues presented throughout this book. Indeed, these reforms are currently underway, and as explained, continue to face a number of challenges. But if their intent is successfully enshrined in statute and is appropriately implemented, these reforms are likely to start a once-in-a-generation shift that will make sound environmental policy, community well-being, and accountable and transparent decision-making, the new trifecta for water access and use in New Zealand.

References

Communities 4 Local Democracy. (2021). Fresh Ideas, Better Water. Retrieved February 1, 2022, from https://www.communities4localdemocracy.co.nz/

Department of Internal Affairs. (2021). Three waters review. Retrieved January 24, 2022, from https://www.dia.govt.nz/Three-waters-review

Department of Internal Affairs. (2022). Three Waters Reform Programme. Retrieved January 14, 2022, from https://www.dia.govt.nz/Three-Waters-Reform-Programme

Gillet, V., Mckay, J., & Keremane, G. (2014). Moving from local to State water governance to resolve a local conflict between irrigated agriculture and commercial forestry in South Australia. *Journal of Hydrology, 519*, 2456–2467.

Jeremiah, F., Suazo, A. E., & Butson, R. (2020). Contemplating the value of liminality for entrepreneurs. *Tamara. Journal for Critical Organization Inquiry, 18*, 1–11.

Ministry for the Environment. (2021). *Exposure Draft—Natural and Built Systems Bill*. Wellington. Retrieved March 8, 2022, from https://environment.govt.nz/assets/publications/Natural-and-Built-Environments-Bill-Exposure-Draft.pdf

Ministry for the Environment. (2022). Overview of the resource management reforms. Retrieved January 14, 2022, from https://environment.govt.nz/what-government-is-doing/key-initiatives/resource-management-system-reform/overview/

Ministry of Health. (2021). Drinking-water. Retrieved March 12, 2022, from https://www.health.govt.nz/our-work/environmental-health/drinking-water

New Zealand Centre for Political Research. (2020). RMA Misrepresentations. Retrieved January 24, 2022, from https://www.nzcpr.com/rma-misrepresentations/

New Zealand Herald. (2018). A consortium of some of New Zealand's largest lobby groups want to see big changes to the RMA. Retrieved January 24, 2022, from https://www.nzherald.co.nz/nz/a-consortium-of-some-of-new-zealands-largest-lobby-groups-want-to-see-big-changes-to-the-rma/NUS7UPAD6L6KC34IOQWMXQKOAE/

Parliamentary Counsel Office. (2022). *Water Services Act 2021*. Wellington.

Söderlund, J., & Borg, E. (2018). Liminality in management and organization studies: Process, position and place: Liminality in management and organization studies. *International Journal of Management Reviews, 20*(4), 880–902.

Stedman, S. J. (1997). Spoiler problems in peace processes. *International Security, 22*(2), 5–53.

Stuff. (2019). National promises to repeal and replace RMA. Retrieved January 24, 2022, from https://www.stuff.co.nz/national/politics/118223167/national-promises-to-repeal-and-replace-rma

Taumata Arowai. (2022). Home | Taumata Arowai. Retrieved January 14, 2022, from https://www.taumataarowai.govt.nz/

Water Industry Commission for Scotland. (2022). Our history. Retrieved March 6, 2022, from https://wics.scot/what-we-do/our-role-scottish-water-industry/our-history

Chapter 10
Conclusions

Abstract This chapter includes final remarks and conclusions based on the elaborations provided throughout the development of this study. It does so by summarizing the book's findings and by reflecting on the current and future status of water-based disputes in New Zealand.

Final Remarks

This study examined the linkage between the commercialization of freshwater and the emergence of hydropolitical conflict. It did so to contribute to the ongoing debate started by environmental security theorists in general and abundance scholars in particular: the connection between natural resource wealth and conflict. In its earliest stages, this study argued that abundance theorists' implicit focus lies in the way in which individuals and groups with competing interests commercialize abundant natural resources and how this commercialization may affect the onset of conflict in turn.

To examine the water commercialization-conflict nexus, this study posed the following question: How and why does the commercialization of freshwater affect the emergence of hydropolitical conflicts? Two central arguments were proposed to answer this question: the first is that the introduction of water commercialization practices influences the emergence of hydropolitical conflicts if they cause economic, environmental, and social incompatibilities for local communities. The second argument asserted that commercial enterprises established through untrustworthy approval and appeals processes were likely to motivate local communities to circumvent the water authority regime and to engage in intense hydropolitical conflicts.

To test the proposed arguments, this study examined the introduction of water bottling and water chlorination in the towns of Ashburton and Glenorchy, New Zealand. The data collected in both locations provides support for the proposed arguments: Conflict intentionality did indeed emerge when respondents perceived that a commercial operation was incompatible with their interests. The data also demonstrated that the containment and intensification of hydropolitical conflicts depended

A. E. Suazo, *The Risk of Water Conflicts in Aotearoa-New Zealand*,
https://doi.org/10.1007/978-3-031-06660-3_10

on the extent to which the respondents trusted the approval and appeals processes behind water bottling and water chlorination. When these practices were approved through transparent and trustworthy processes, they were likely to motivate participants to engage in intra-institutional, low-intensity conflict, thus allowing water authorities to contain conflict at the lowest possible levels. In contrast, when these practices were approved under conditions that respondents did not regard as legitimate and when their appeals processes did not allow for the adequate ventilation of concerns, some members of the above communities were inclined to circumvent the dispute resolution mechanisms available through their water authority system and likely to espouse extra-institutional, high-intensity conflict intentionality.

This study also offered four different reflections of its limitations. First, it emphasized the explicit focus on rural communities and argued that in order to attain a clearer, country-wide perspective over the water commercialization-conflict nexus, research needs to adequately examine how these dynamics unfold in urban environments, where conflicts over different water practices continue to emerge and intensify. Second, it recommended a special focus on the study of hydropolitical conflicts as they apply in Māori communities, as the generation of water incompatibilities, and Indigenous communities' distinct views on water governance, may lead to different outcomes in these contexts. Third, in stressing the empirical benefits of examining water bottling and water chlorination, this study also recommended that attention needs to be devoted to the interlink between dairy farming and the emergence and intensification of conflict, given the dairy industry's reported role as a source of frictions within and without communities in New Zealand. Lastly, and akin to the previous recommendation, this study presented a brief case for the future study of water securitization strategies as causal determinants of hydropolitical conflicts in New Zealand, attempting to examine how and why water-based conflicts could emerge in view of continued pressures to securitize the country's freshwater via treatment.

Beyond the empirical findings in this book, intersections were made with a series of proposed reforms to New Zealand's Three Waters and Resource Management systems. It was found that the policy intent behind these reform programs is likely to motivate improvements to how hydropolitical conflicts emerge and intensify. These programs afford novel opportunities to contain water-based disputes within the institutional framework provided by the water authority system, thus reducing the risk for water conflict intensification. Challenges however remain in the introduction of these changes, which require acknowledgment of the cognitive processes of change, and how they impact communities' likelihood to embrace and challenge a shifting condition.

The Coleridge Dilemma and New Zealand's Future

Life has taken me on a path where I have had to settle and resettle in seven countries in the Americas, Europe, and Oceania. I consider that no other country in which I have lived bears the promise that New Zealand harbors within its shores. The land of the long white cloud so happens to also be the land of dialog, reconciliation, and remediation. I am aware that this book touches upon very sensitive subjects for many communities across New Zealand and that it addresses them at a time of significant change for the country. Beyond its theoretical and empirical merits, I hope this book contributes to something I consider to be much brighter: Despite conflicts over water, the opportunity to dialog is still strong in New Zealand.

The Coleridge Dilemma was proposed earlier to describe the dichotomy between water abundance and the emergence of conflicts over said abundant water. It does not however fully encapsulate the New Zealand psyche of togetherness and ingenuity. Indeed, there has been a recognition that water management has not been optimal, but alongside this admonishment, there has been swift action. Whether this action came in the form of unilateral community-based solutions, art exhibits, or through the development of significant reforms to the country's water authority and resource management systems, there is little doubt that New Zealanders see in adversity an opportunity for positive change and that they are not shy to pursue these changes. They approach these changes despite not always receiving the domestic and international recognition these actions deserve. I hope this book contributes to acknowledging and paying tribute to the spirit that drives positive change in New Zealand in all its forms.

Appendix
Questionnaire Guide

Value

1. How would you describe your satisfaction with the quality and quantity of water in your community?
2. What does water mean to you personally? [For recreational use? For domestic use? For profit?]
3. How is water related to your occupation? [Does your business or place of employment depend on water for its activities?]
4. In your view, is your business or place of employment's water access under threat?
5. Who owns the water in your community? Does everyone have the right to it?
6. Who makes decisions over water in your community?
7. Are you satisfied with how decisions over water are made?
8. There are (or were in the case of Ashburton) plans to bottle water from your community. How do you feel about that? [Should people or companies be allowed to sell water?]
9. What do you think about water being bottled or sold elsewhere in New Zealand? Do you find it problematic?
10. Public water supplies in New Zealand are consistently being treated to ensure they are safe to drink and use. How do you feel about your public water supplies being treated?
11. In situations where, for any reason, water is not safe for drinking, what should the authorities do to improve the situation in your view?
12. In descending order (from 6 to 1), what activities seem most important to you:

    (a) Water Bottling
    (b) Irrigation
    (c) Rafting
    (d) Water Shipping
    (e) Fish Farming

A. E. Suazo, *The Risk of Water Conflicts in Aotearoa-New Zealand*,
https://doi.org/10.1007/978-3-031-06660-3

(f) Domestic Water Use.

Perceptions of Abundance

13. In your opinion, do you consider New Zealand to be a water-abundant country? Can you elaborate? Where can we see that abundance?
14. In what ways do you (and/or your community) benefit from this water abundance?
15. In your view, is the abundance of water in your community under threat in any way?

Perceptions of Water Asymmetries

16. In your view, do people benefit from water in the same way?
17. Do people in your community use more water than others? Is this a problem to you?
18. Do you think your own access to water is at risk? Why?
19. Are you aware of anyone in your community having trouble accessing water?

Indicators of Conflict Intentionality

20. If your water access was reduced by someone withdrawing water for sale, would you consider this to be:

    (a) Unfair
    (b) Illegitimate
    (c) An issue worth resisting
    (d) Acceptable.

21. Under what conditions would you decide to take political action if you had a concern about water?
22. In connection to the previous question, what type of action would you consider taking:

    (a) Protest, Vandalism
    (b) Litigation
    (c) Pushing for fines and permit denials
    (d) Petitioning
    (e) Voicing opposition, writing editorials
    (f) None.

General

23. Would you consider Ashburton/Glenorchy to be your home? [How long have you lived in the area?]
24. Are you employed in the area?
25. What do you do for a living?
26. DOB
27. Gender
28. Ethnicity

29. Age
30. Political affiliation.